ANCIENT ENERGY

Key to the Universe

Published in San Francisco by

HARPER & ROW, PUBLISHERS, INC.

New York, Hagerstown, San Francisco, London

ANCIENT ENERGY
Key to the Universe

Maxine Asher

For information address Harper & Row, Publishers, Inc., 10 East 53rd Street, New York, N.Y. 10022. Published simultaneously in Canada by Fitzhenry & Whiteside Limited, Toronto.

FIRST EDITION

Designed by Patricia Girvin Dunbar

Library of Congress Cataloging in Publication Data

Asher, Maxine.
Ancient energy.

Bibliography: p. 176
Includes index.
1. Occult sciences. 2. Geomancy—Miscellanea. 3. Man, Prehistoric—Miscellanea. 4. Antiquities—Miscellanea. 5. Curiosities and wonders—Miscellanea.
I Title.
BF1999.A68 001.9 78-19497
ISBN 0-06-060398-9

79 80 81 82 83 10 9 8 7 6 5 4 3 2 1

Contents

Foreword

by Lowell Ponte

Walking along a lonely beach a native finds a radio washed ashore with the seaweed. He has never before seen such a thing. Cautiously he picks it up, touches parts of it that move. Intermittent noises crackle from it, and in panic the native drops it. The noises stop. For the rest of his life he will tell others of the talking box, but few will believe him. Such things are impossible except for those rapt with illusion or delusion. Shunned, the native retires to a solitary cave with his box. Sometimes when he shakes it the sounds return for a moment or two, but none will hear because none will listen.

Dr. Maxine Asher is a beachcomber on one of the farther shores of scientific exploration, and in her view we are the natives of this story, not the sophisticates. Ancient peoples built their temples and pyramids using a logic we no longer understand, she believes—a logic calculated to tap subtle energies in the environment. Why else would peoples all over the world have moved heavy stones great distances to Stonehenge or Giza with little apparent reason—if not to use energies they felt in such places?

What, precisely, these energies are is uncertain. But Dr. Asher and many other people think they have felt their effects in a host of physical symptoms and intuitions. Dream intuitions led the explorer Heinrich Schleimann to discover the buried ruins of ancient Troy, once thought of only as a fictional city in Homer's *Iliad* and *Odyssey*. Similar intuitions have, during the past 20 years, impelled Dr. Asher into an odyssey of her own, a search for links among ancient civilizations that has taken her from Egypt to Ecuador, from Malta to Mexico, from Arizona to Africa and beyond, across five continents with a tireless energy the ancients might well have envied.

I first met Dr. Asher in 1974 when assigned by the Public Broadcasting Service to do a television interview with her. I have since come to know her as a friend and neighbor and have watched with fascination as she has tried to hold to a delicate beachhead with dogmatic scientists to the right of her and fanatic occultists to the left. As a science writer, I can appreciate the challenge she faces, and the courage she shows, in daring these quicksands of controversy in her search for hard evidence. It was just such courage and determination by earlier scientific pioneers that discovered for us all the pearl of astronomy in the oyster of astrology and the essence of chemistry in the murk of alchemy. Doubtless there are many other kinds of valuable knowledge that can be sifted from primitive information and ideas.

To make such discoveries, Dr. Asher believes, we must use our *whole* brain—not only the left hemisphere, with its skill at rationality (and rationalization), but also the right hemisphere, with its grasp of proportion, dimension, color, texture, and all the other perceptions we try to tag with quicksilver words like intuition, creativity, imagination. Why were the ancient Greeks and Egyptians so concerned with exact proportions in their sacred buildings? Why, in English, do we speak of an insane action as being "without rhyme or reason," as if the quality of coincidence or synchronicity in "rhyme" could justify an action. The poet is as legitimate, by this standard, as the scientist. Both have uniquely valuable perceptions. Many scientists now walk barefoot on the farther shores of knowledge, and Dr. Asher is among them. Can they explain anti-matter? Or forces that appear to be moving backwards in time? Or the reason why a crystal in a radio can capture resonant energies transmitted thousands of miles away? As yet, these scholar-pioneers lack a firm footing, but they dare to try new ways and means to find answers.

Soviet scientists apparently think something is afoot with ancient energies. By charting lines among ancient sites, they have evolved a theory that our planet functions as one giant crystal focussing electromagnetic and other energies along its facets. Should we be concerned that the Soviets have set up a transmitter complex, designed to alter Earth's magnetic field, on precisely the longitude of the pyramid at Giza and another on precisely the longitude of Salonika, Greece, shaken by devastating earthquakes in 1978? As a former Pentagon specialist on such exotic weapons technologies, I think we can ignore such ideas only at our peril, however controversial they might be. Dr. Asher's book provides a heady dose of such food for thought. May "the force" be with her.

Lowell Ponte
Santa Monica, California
August 1978

Acknowledgments

No man is your enemy, no man is your friend, every man is your teacher.

I would like to state my appreciation and unqualified gratitude to the following people, who gave me their moral, spiritual, and intellectual support during the writing of this book: my mother, Charlotte Klein Jamison, and my daughters, Laurie and Jan Broslow and Susan Moore; Dr. Mildred Murry; Lowell Ponte; Jane Fultz; Dr. Rhoda Freeman; Meg Downs, Earl and Jane McNevin, Lloyd Daughenbaugh; Dr. Michael Hughes; Frank and Rita Stricklin; Bella Karish; Mitch Uribe; Moira Carney; Tony and Sherry Arlotta; Jaine Smith, Carmen Barrios; my editors at Harper & Row, Ann Bradley and Marie Cantlon; Paul Horn; Ann Miller; Francisco Salazar Casero; Raul Mendoza; Suzanne Crayson, Grant Gudmundsen, Pauline Mann; Marvin A. Luckerman; and Aboul Magd Mahmoud. Above all, I would like to thank the friends and members of the Ancient Mediterranean Research Association who continually supplied me with the energy to tackle the difficult task of writing about the unknown.

1. The Case for Ancient Energy

There is a power in the universe, a mysterious energy, that can transform your life. This force, which I call ancient energy, can provide you with a key to unlock the mysteries of your own origins and the secrets of creation. What is more exciting, you can learn to tap into this force at will, to make it work for you in dozens of ways just as the ancient people did thousands of years ago. To experience ancient energy is to stand in awe before the wonder of the cosmos. Nine years ago I walked into a strange, new, and powerful world that changed my life and provided me with knowledge not attainable in any other way. This book tells how it happened and how it continues to happen in my own life and the lives of my colleagues who dare to explore its potent challenge.

I had been a studious child who developed into an overeducated adult. For me, knowledge was garnered from textbooks, college outlines, and austere lectures. If something wasn't printed or spoken, it didn't exist. My avid interest in archaeology began after graduation from college and took me abroad where I travelled extensively to research mankind's earliest origins. My penchant for learning was scholarly, and I scoffed at theories that proposed ideas I could not immediately see, hear, or touch. By age twenty-seven, I had earned six degrees and teaching credentials, was married, had two children, and was comfortably settled in suburban California where I held an elementary school teaching position. But as a result of several strange events that took place in the spring of 1968, my life was destined to change. I began to have clairvoyant dreams—dreams of situations that later happened in the real world—usually within days after my premonitions. These dreams, to which initially I gave little credence, were sometimes warnings about accidents or illness but most of the time they portrayed some joyful event that was about to happen to me or to my family. Occasionally I had advance notice of money which came unexpectedly in the mail, and usually I could predict what would happen at home, at school, or on vacations. These detailed previews of the future interfered with the normally rational way I viewed my life and were complicated by the reaction of my husband, who called me the "house witch." Naturally, I tried to find a scientific explanation for what was happening. My search began in the neighborhood libraries, where I scanned books, articles, and other references to accounts about mysterious experiences like mine. The rationale given in the texts for clairvoyance was vague; at best it suggested that people who reported such mystical events had psychic or intuitive powers, but little convincing data to support such ideas existed. I dismissed these written accounts as the ramblings of fictional writers who had reported the same kinds of happenings throughout the history of the human race.

In the meantime, I pursued travel and explorations abroad, where I investigated temples, tombs, and monuments of the ancients with a kind of grand passion to understand more about their origins and their creators. During the summer of 1969 my search took me to Europe. I had long since discounted the dream phenomena of the previous year. In fact, I attributed my experiences to overwork and a badly needed holiday. It was at this time I had my first encounter with the force I now call ancient energy. It happened over a series of three days while exploring the tombs of the Etruscans in Cerveteri, Italy, a tiny city northwest of Rome in an area known as Tuscany. I knew very little about the Etruscans; they had always been an enigma to historians. Where they came from and the meaning of their writings were unknown, but they excelled in the areas of art,

Ancient energy may be heavily concentrated at this Etruscan grave site near Cerveteri Italy.

religious ritual, and athletics. Usually, my husband accompanied me to the sites but, on this occasion, he was exhausted from the long overseas trip and preferred to remain at the hotel. So, map in hand, I proceeded alone to the burial grounds.

I entered the first of the tumuli (the name given to the rounded domes covering the tombs), and sat down on a cold stone slab with my notebook and pencil in hand to record my observations. Fortunately, I was alone and could take in the entire chamber with a sweep of my eye. Suddenly, I was drawn to a sarcophagus far back in the room; on its lid was a statue of a reclining Etruscan with his eyes wide open and staring. I had never seen eyes like that before on a statue. They had no eyeballs or inner construction, just sockets, but I felt a connection between my eyes and the orifices like a lightning bolt between two steel rods. I must have stared at the statue for fully five minutes. Then, looking away, I broke the current and sat down on a fluted pillar next to the casket. I opened my notebook and fifteen minutes later had written pages of information about the Etruscans, their society, their accomplishments, and myriad other details. I wrote as though taking notes from an invisible lecturer; I had never consciously learned the information which filled the tablet. When I read back over the notes, they seemed strange and unfamiliar. The next day I purchased some books on Etruscology, and little by little I verified most of what I had written in the tomb. What puzzled me was certain bits of information that I could not verify anywhere and I knew I had not read or learned about in school. I reasoned that perhaps I had acquired the data somewhere in the past and had forgotten it or that what I had written was some strange figment of my imagination.

Anxious to repeat the experience, I returned to the tombs the next day but nothing significant happened. They were filled with tourists and I felt uncomfortable about conducting any further observations. After I left the chambers, I decided to visit a small museum located near the site. I told the curator that I was studying to be an anthropologist, and he was more than happy to allow me to hold several pieces of Etruscan sculpture. Immediately on contact with the artifacts, I felt a current run through my body. The reaction was overwhelming. I was afraid I might drop and break the precious statues. One figurine was comprised of a man and woman pulling what looked like an oxcart. In my transfixed state, I could mentally picture an entire scene of Etruscan agriculture taking place thousands of years ago. The curator was explaining the piece to me in his broken English. But of course he didn't have to. I already knew everything he was saying. I even knew where the sculpture had been found before he told me.

This experience was so shattering I couldn't even feel much joy over the new knowledge I had acquired because I was more concerned at my inability to understand the strange way in which it had come to me. That night I tried to relate two days of events to my husband in our hotel room, but I noticed immediately that he thought I had gone mad. This was the first time I realized how difficult it was to explain such experiences to others. His skepticism indicated that I had better keep my thoughts on the subject to myself until I could begin to understand and describe what I was observing as reality. Now, analyzing the events after eight years, I realize that the ancient sites conducted an energy which affected my consciousness in a powerful but fascinating way.

Immediately after the encounter in the Etruscan tombs, I began searching among my colleagues for those who had undergone similar events, to reassure myself I was not alone in my reactions. The first indication that other intellectual people shared such perceptions occurred when I interviewed a professor of geology from Oregon to secure data for my doctoral dissertation entitled, "Recent Theories of Intuitive Perception Applied to Ancient Anthropological Inquiry." I had selected ten professors in various academic disciplines for the oral interview section of the paper and had prepared a list of questions for them in order to find out if they used intuitive data as valid information in their scientific research projects. The professor and I met at the Los Angeles International Airport, and he offered to take me to dinner since he was anxious to share his experiences about a recent undersea project near the Azores in which he had taken part. When I attempted to give him the standard questions slated for the interview, he tried to dodge the subject. So we continued talking about his work and also my strange encounters with ancient energy at the archaeological sites. I told him how I had been drawn to markings on a pathway at Gibraltar which turned out to be actual prehistoric inscriptions and explained my other forays into caves and remote areas where I felt drawn to artifacts and symbols that had not been explored before. After several hours we had finished two large bottles of wine, and he seemed eager to talk. What he said startled me.

"Listen," he confided, "I've never told this to anyone before but I must tell you what happened to me in South America." He went on to describe his reactions at Machu Picchu, an Inca fortress high up in the Andes where he had been one summer morning at dawn. The illustrious doctor described how he had lain down momentarily on a flattened area of standing stones when suddenly he heard music and saw Inca ceremonies. At first he thought he must be dreaming but quickly realized he was wide awake and consciously aware the experience was completely real. He was so alarmed by the event he left the site immediately and returned to the United States the same day. I reassured him that I had had similar experiences in my own travels, and he was somewhat comforted. For the rest of the evening, we exchanged dialogue, trying to seek some orderly explanation for our common experiences at the ancient sites.

Up to this point my own reaction to ancient energy had been pleasurable. At the least, my exposures to the force had been beneficial, enhancing my knowledge of prehistory and energizing me to the point where I was able to walk up to twenty miles a day between locations. In Los Angeles, I was usually exhausted after walking two blocks, but after exposure to the sites, I frequently had a boundless energy that kept me going long after my normal physical limitations would have been reached. I also discovered that my experiences were more penetrating if I dressed in simple light clothing and carried few objects or possessions such as purses, duffel bags, and so on. The more I encumbered myself, the less effect the site appeared to have. I realized later that too much clothing cut down my ability to receive the energy. In addition, the impact was considerably lessened when my husband accompanied me and totally deadened if several people were at the site. Shortly after the Etruscan experience I began to travel alone in order to maximize my sensitivity to the force. Often it seemed that when I was surrounded by a group, the effect was like several radios broadcasting different stations at the same time. I believe this interference occurred because each individual simultaneously transmitted and received the energies at different levels of consciousness. The accidental result of such disharmonious wavelengths counteracted my own ability to contact the force. This made me suspect that humans may unconsciously transmit a flow which is picked up by other receptive individuals. The vibrations may at times, enhance or deplete another receiver of energy depending on his or her ability to conduct and release the charge each is receiving. While animals appear to let the forces flow through them naturally, discharging vibrations routinely as they move in harmony with the normal rhythm of the universe, humans sometimes run into trouble since they try to control or stop vibrations. I believe many people in contemporary society are uncomfortable with the conflict between natural energies and the vibrations produced by the strains of urban life which frequently work against the laws of nature. To survive the disharmony, which has come to be a major factor in our modern technological world, people often become engaged in a constant round of activity or else they suppress the flow, sometimes in unhealthy or unnatural ways. Such ideas were new to my own way of viewing the universe, but I continued to grope for answers about the unknown energies since I was fascinated by the possibilities they held for acquiring new knowledge.

My excitement over the phenomena of ancient energy was shortlived. I soon discovered that this strange power also had dangerous aspects, which prompted me to treat it with great respect in my subsequent travels. My caution was due to an alarming incident that occurred in 1975 as I was exploring the Yucatan Peninsula of Mexico with a fellow researcher, Pauline Mann.

I had been warned not to go to Yucatan by my mother and brother, both of whom had had dreams and premonitions that I would be killed if I went. I disregarded such warnings since my own intuition indicated there was no cause for concern. Pauline and I were scheduled to leave Merida at 10 A.M. and travel by car to the great pyramid at Chichen Itza, a sacrificial Mayan monument some two hours drive from the city. I felt refreshed and alive when we departed for the exploration, as I always did in Mexico. We talked animatedly for the first hour and I was laughing about the warnings I had received about not taking this trip. About three-quarters of an hour from our destination, my head began to ache. I grabbed two aspirin, gulped them down with some bottled water we were carrying, and ignored the discomfort. About fifteen minutes later I started to feel weak and broke out in a cold sweat. "I don't know what's happening to me," I said to Pauline, "but I feel like the blood is being drained out of my entire body."

Pauline stopped the car at the side of the road and looked at me. I was slumped over the seat and my face was chalk white. We were twenty minutes from Chichen Itza. I said, "Get me to the nearest hospital. I'm going to die." She said, "You won't die. I'll drive faster. Just close your eyes and relax." I vaguely remember her saying, "We're passing Chichen now," and then I have little recall of the time that elapsed. All I know is that Pauline did not stop at the nearest hospital which she felt would not give me prompt and adequate treatment. She simply continued driving toward the east coast, ending up at Tulum, a site I also had been longing to see. We arrived just twenty minutes before the gates were to close for the night. I had already recovered and was full of energy. The experience in the car felt unreal and I was relieved that it had passed. That night I had dinner on Isla Mujeres, near Tulum, as if nothing had happened. I felt wonderful and the whole scene reminded me of my earlier precognitive experiences when I sometimes dreamed about a frightening event and later awakened feeling a kind of joy that it was over. Yet, in this instance, Pauline witnessed that it had really taken place. What was responsible? Could it have been ancient energy? Was the same positive experience from the Etruscan tombs related to the frightening manifestations at Chichen? I thought this was possible. But why? Why would one site give me energy while another had adverse effects? Why was it difficult to recall these experiences after they happened and to explain them in concrete terms? I had no answers, but I was prepared to conduct an inquiry into the phenomenon in order to gain new understandings so I could explain the strange events to myself and to others. Most of all, I sensed that the essence of ancient energy might be important in my investigations of the ancient world since I had been fascinated by certain unresolved problems of antiquity like the possible existence of Atlantis, the construction of the great Egyptian pyramids, and the unknown builders of the giant megalithic monuments at sites such as Stonehenge and Machu Picchu. I knew from ancient texts, for example, the Dead Sea Scrolls and the Nag Hammadi Codices, that ancient people were involved in mysticism, secret doctrine, symbology, and cultism. It was this facet of prehistory which held my interest and which seemed, in some remote way, to be connected to my experiences with ancient energy at the archaeological sites.

Some years passed before I could find a satisfactory explanation for the new force I had identified but which still did not lend itself to any precise definition. My inquiries resulted in dozens more visits to the ancient sites, the recording of experiences as they happened, and extensive reading. Many years after the Etruscan encounters, I was cautiously optimistic that I had some understanding of the problem. Yet certain basic questions still remained to be answered. Where does ancient energy come from? How does it affect human beings and in what manner is it conducted? What precautions have to be taken when experimenting with the force? And how could humans use it to achieve the greatest effect? These questions had no definitive answers no matter how much data I gathered, an understandable dilemma in view of the embryonic nature of such a quest. All I could do was experience, observe, and record information based on my own encounters and the reports of a few other researchers like Dr. Jeffrey Goodman and Dr. Thelma Moss who were groping with problems similar to my own. The scope of the energy question was immense and the reliable data both elusive and scarce. Yet ultimately, given time and the patience of a few dedicated explorers in the field, the development of a new system of scientific methodology was possible which could open the way for new modes of thinking about matter and energy.

One positive result stemming from my investigations has been the development of a working definition of ancient energy. Let me share it with you. I sense that ancient energy is an invisible force composed of matter vibrating at different wavelengths which is governed by the laws of quantum mechanics, the only system of measurement which seems to me most applicable to the experience I had encountered. The quantum system is another way to explain the action of elements by using various mathematical equations to refer to particles and waves. In such a system, wavelengths may act as particles and conversely. Quantized particles do not abide by the rules of time or space but have within them a span of life ranging from a fraction of a second to millions of years. Ancient energy may be that part of the quantum wave composed of particles with life spans long enough to have endured from earliest times to our present age.

Thus, an ancient energy mass, initiated in our prehistory, remains in force because it has a more enduring structure in the quantum scheme. It is my opinion that ancient energy particles are actualized through the intervention of human consciousness coupled with a faith or belief that such energy exists. The more expanded the perception of the receiver, and the stronger the charge to which he or she is exposed, the greater the effect ancient energy will have in terms of a physical manifestation. I also believe we receive ancient energy through our normal five senses and through an expanded process of awareness which I shall describe later in detail.

Since ancient energy appears to be governed by principles which do not abide by the cause and effect system we normally apply to the physical universe, the events I have experienced lead me to believe that the energy mass or the individual particles can alter their rate of motion, life span, and intensity depending on the action of human consciousness, or as a result of structures built in certain geometric forms, such as the pyramid, dome, or hexagon. I have observed how the interior of the Great Pyramid of Cheops caused an altered state of awareness among participants on two of my expeditions to Egypt, and I have witnessed the action of certain sacred sites in the Yucatan as my fellow researchers interacted with the heightened energy forces I believe exist in such areas. As I stated earlier in my working definition, human consciousness and a certain belief system are necessary for the activation of the force. Therefore, not everyone on my expeditions was affected in the same way and some were not aware of ancient energy at all. However, the incidents that occurred over the many years became so frequent and had so many elements in common, that my interest in finding a rationale behind the phenomena increased. I believe that some prehistoric cultures had a natural understanding about the essence of ancient energy and used its power in many daily endeavors including megalithic construction, the healing arts, and nonverbal communication.

Corroboration of my own research was more difficult. However, I discovered that a few limited experiments with ancient energy had been conducted both at early human sites and apart from them. For example, Jeffrey Goodman, in *Psychic Archaeology*, describes archaeological discoveries based on the energies received and reported by psychic informants at a Flagstaff Arizona dig during 1976. In England, Paul Devereux, publisher of the *Ley Hunter,* devised instruments to measure the flow of energy at major megalithic sites. Similar work is described by Francis Hitching in his book *Earth Magic*. Frank Waters in *Masked Gods* documents the use of such energy in Navaho and Hopi healing techniques. In all cases, the phenomena were not termed ancient energy but the characteristics appeared to be identical to the force I had identified.

Researchers like Peter Kolosimo and Andrew Tomas refer to the phenomenon as cosmic, universal, or timeless energy but have not described it in detail because of its elusive and complicated nature. Although the energy itself appears to have some of the same effects as descriptions of psychic power, the two should not be confused. Psychic response is a second way of knowing or perceiving through the right hemisphere of the brain as distinguished from the left hemisphere where physical data from the five main sensory organs is received and translated into rational thought processes. Psychic ability operates from the right hemisphere of the brain and processes information secured from human sensing devices beyond the range of the five traditional senses. Data entering the right brain is holistic—that is, the form in which it is received is essentially conceptual or symbolic as opposed to the myriad pieces of physical information which enter the left brain. However, it is the left brain which must give meaning and significance to the right brain data. Although psychologists know this process takes place, no one is certain precisely how the transmission and evaluation of concepts is translated by the left brain into the physical reality. Right brain-left brain function has been reported extensively by Robert Ornstein and is still a prime focus for study by psychologists. I believe that ancient energy is a mass of stimuli which is sent to the right brain only under certain specified conditions. These are: the intensity of exposure to the energies, the human state of consciousness, proper conduction, and a requisite belief system. The latter operate beyond the range of our traditional means of sensing and perceiving and are therefore very difficult to isolate, identify, or define.

Often, ancient energy appeared to affect me in a more

The author examines a prehistoric site in Limerick Ireland using traditional archaeological methods and intuitive clues.

powerful way when I consciously blocked my five main senses or had them dulled as a result of fatigue or illness. During these periods of repressed sensation, other perceptual centers apparently became keener, particularly those receptors involved in the conduction of ancient energy. This effect can be compared to some blind people whose hearing is exceptionally acute as a result of the impairment of the visual sense. It also can be likened to the sleep state when sight, hearing, touch, and smell are minimized and the brain's unconscious centers manifest the data they receive into visions and dreams. I also experienced ancient energy more keenly when travelling abroad and visiting archaeological sites even though my early encounters occurred at home in California. When my husband failed to understand why I was constantly away from home in my quest to learn more about ancient energy, I explained the importance of the foreign centers in my investigations since experience with different and more powerful energy fields was a key factor in the research. Since I had spent most of my life closed off to concepts related to the energies, the first experiences with the forces were accidental. In fact, I was barely open to such concepts when I began my European investigations, and I believe the sites were largely responsible for awakening me to the reality of ancient energy. I feel sure that my understanding of the forces, and my developing consciousness in relation to them, would have been curtailed or even stopped in the earlier stages of my awakening if I had not experienced the feeling of total immersion at the archaeological sites of the Mediterranean and the American Southwest.

During group investigations, the most resistant subjects, when surrounded by high energy fields, opened to the possibility that such phenomena actually existed. In such high vibrational sites, it appeared that no one resisted or blocked the forces since the feeling they engendered was one of joy or exhilaration. For this reason, I feel that it is essential for people interested in experiencing ancient energy to personally visit the sites, at least in the early phases of their investigation. Awareness of ancient energy can sometimes be accidental but developing a continual sensitivity to it requires training and multiple experiences on the part of the seeker. If more subtle levels of consciousness are never opened, which is true for the vast majority of people, the subject may be surrounded with the forces and still not recognize or even acknowledge their power since the experience is not part of one's accepted belief system. The first encounter is the most crucial one. After the initial breakthrough, people generally know what to expect. I have seen individuals at first ignore cues that would normally stimulate my own perception and direct me toward a discovery of some new knowledge. Later they became expert at recognizing intuitive clues and became active participants in the research.

My own inquiries resulted in the formation of several theories that may explain the origins of ancient energy. One theory traces the beginnings of the force to "outer space" which term, in its broadest sense, refers to dimensions beyond our normal sensory plane. Outer space is believed by Dr. William Tiller of Stanford University to coexist with the physical dimension we normally perceive, and the two systems are frequently thought to operate side by side. Dr. Tiller suggested this theory in a speech given before the International Yoga Festival conference held in Phoenix, Arizona in February 1977. Tests to measure the effect of this energy on human consciousness have been conducted with sensitive mechanical devices like biofeedback equipment but more frequently they manifest in the dream state or as precognition, reported by subjects with more finely tuned sensory processes. They were reported by astronaut Ed Mitchell during space travel, where he claims that cosmic vibrations exerted a powerful effect on his consciousness. In his book *Psychic Exploration—A Challenge for Science*, Mitchell states that the properties of space caused him to receive certain knowledge "noetically" as an experiential cognition gained through private subjective awareness. Such perception, states Mitchell, is as plausible as the navigational programs and communications systems by which the space ship was operated. Arthur Koestler refers to a particle of matter called the neutrino which he suspects is governed by the quantum energy equation. In *The Roots of Coincidence* he observes that the neutrino could be the basic transmitting entity of the telepathic response. Since neutrinos, like other quantized particles,do not abide by the rules of time and space, I believe they may be the agent which carries prehistoric communications from the ancient sites, which are channeled through consciousness to visitors at the sites, assuming that the basic conditions for the reception of ancient energy are in force. Thus, certain locations may still hold messages from their builders which are transmitted to sensitive individuals visiting the areas. My "knowing" the exact place where the Etruscan artifact was found could have been an example of this kind of ancient energy transfer. The knowledge which led to the construction of megalithic sites, the development of ancient astronomy and other sophisticated sciences, and the advancements made by early man in the healing arts did not come from sources currently known to archaeologists. Instead, modern researchers are puzzled by the technical expertise possessed by the ancients and the fact that they knew exactly how to live within a hostile environment by maximizing and using the forces of nature to their advantage. It is my opinion that they received their knowledge by means of messages carried by the neutrinos which may have been enhanced by the power of the sites themselves. Such knowledge was probably received automatically without conscious awareness by the receiver. Prehistoric man simply had an immediate knowing of how to proceed in a variety of practical endeavors. In *The Brain Revolution*, Marilyn Ferguson identifies at least twenty receptors of energy in the human body, including sensitivity by some people to free electromagnetic energy, sonar-like capabilities, and special kinesthetic skills. Often, while our senses are still firmly grounded in what we consider to be physical reality, supersensory mechanisms are simultaneously in operation governed by the quantum laws. This leads to a "spaced out" feeling which is possible to experience without taking drugs as a result of extrasensory receptors tuned in to energies the brain is not accustomed to receive. It is possible that ancient people never learned to block their extended modes of perception and were more finely tuned to the higher wavelengths in the environment.

In *Mystery of the Mind*, Dr. Wilder Penfield describes a mind energy, generated by certain highly developed individuals, which can only be created while the physical brain is still alive. The energy transmitted by such aware humans does not disintegrate, according to Dr. Penfield, but remains in the universe where it can be contacted by other receptive "intelligences" or by humans sensitive to the vibrations. Dr. Penfield takes great care to distinguish mind energy, which is eternal, from brain energy, which is temporal. He believes that the development of mind energy is a sort of assurance that the human race can achieve immortality. If this theory is correct, I believe the ancients may have generated a mind energy which can still be contacted in the environment of the prehistoric sites and also away from them. Since the phenomena described appear to have similar characteristics, Dr. Penfield may be using the term mind energy in the same way that I am using ancient energy.

The inertia provided by exposure to ancient energy does not automatically lead to eternal life. However, I must confess feeling a sort of immortality whenever I visited the sites. Occasionally, I lost all track of time and a sense of knowing precisely where I was. These feelings were very apparent in Ireland and Spain, two countries which I believe are saturated with an ancient energy mass. Although the fields of force transmitted a positive charge, giving me a feeling of great exhilaration and joy, I frequently felt drained of energy within hours after leaving the sites, almost as if the vibrations which had kept me aloft were suddenly drawn away. This "deflated" sensation became even worse when I moved completely out of the area sometimes leaving the country by plane within a short time after exploring the prehistoric ruins.

The belief system necessary for the activation of ancient energy brings a certain religious component into the scope of scientific inquiry perhaps for the first time in the history of intellectual thought. Science and religion have, by tradition, remained in separate camps. Throughout man's existence on this planet, one or the other philosophy has held sway. When the theories of Copernicus overthrew church doctrine in the Middle Ages, the interests of the scientific community became the dominating force. Although religious thought related to scientific problems never regained stature, in the last decade a few academics have begun to examine spiritualism with an intellectual approach, especially as it applies to questions of energy and matter. Dr. William Tiller suggests that the spiritual essence flows in what he calls "negative space time" and does not become actualized until some synchronous occurrence on earth gives it physical time and place. In other words, "negative space time" is another dimension of the universe which exists as a mirror image of our own reality, performing the exact opposite of everything we perceive in our own plane. In the case of the spiritual essence itself, realization that such phenomena exist depends on the belief system of the searcher and the degree to which his consciousness is developed in the direction of perceiving the higher energies. This idea, applied to ancient energy, led me to believe that the fervent religious life of ancient people actually helped to sharpen their perception so that they could receive energy operating beyond the range of their five senses. Since the ancient belief system was already strong, prehistoric man had all the components to help him maximize the force and channel the energies into advanced feats of construction and technical accomplishment. Ancient sites, like Stonehenge, became very powerful because more than one culture, I believe, created a field of force at the same location. At least nine layers of cultures were found buried at Troy in Turkey and at Stonehenge, according to Geoffrey Bibby, in *The Testimony of the Spade,* there are layers of monolithic monuments buried under the present structure.

We cannot be certain exactly how the ancients channeled the energies so that they produced physical results. However several authors have alluded to the possibility that levitation was a process achieved by the ancients. Certain highly spiritual sects in Tibet actually claim to perform extraordinary acts by which humans can lift themselves from the ground and stay aloft for periods of time without material assistance.

In *We Are Not the First* Andrew Tomas suggests that the blocks comprising the Great Pyramid at Giza may have been lifted up and set into place through levitation, a process which could have rendered the blocks weightless. After personally examining the construction of the pyramid, I think his theory has interesting possibilities for future research especially in light of levitation experiments successfully conducted by the Russians and documented in *Psychic Discoveries Behind the Iron Curtain* by Ostrander and Schroeder. In the Russian experiments, the objects moved by means of concentrated mind energy were small in size, perhaps two inches in diameter. However, the implication was that collective levitation might have the potential to move larger articles. Warren Smith reports in *Lost Cities of the Ancients Unearthed* that the great stones at Tiahuanaco in Bolivia may have been transported in the same way. My own theory, based on observations at more than a dozen megalithic sites, is that some form of energy, transmitted by the ancients, was a major factor in the construction of the monuments and may also account for the fact that so many of the megaliths are standing intact after thousands of years with not one drop of mortar or other adhesive to hold the blocks together. In February 1978, a Japanese archaeological team from Waseda University near Tokyo tried to reconstruct a pyramid exactly like the structures at Giza near Cairo. Their attempt proved to be completely unsuccessful since they could not "master" the techniques of building used by the ancient Egyptians. After witnessing their futile efforts, the Egyptian government ordered the project halted and destroyed, midway through the enormous effort.

The ancient tombs, temples, pyramids, and megaliths probably had a unique purpose in the lives of early man aside from religious and ceremonial functions. I believe they may have served as giant power plants which concentrated and focussed energies because of their unique forms. Perhaps these structures also made use of energy that is believed to reside beneath the earth's surface, and they also may have responded to the humans who built and inhabited them by forming a kind of circuit which assisted in healing, food preservation, acquisition of certain scientific expertise, and the creation of an atmosphere of well being or protection from the harsh outer environment. All we know about the ancient world is that such feats were accomplished by early man, but we cannot be certain how they came about. Certainly they have not been replicated in modern times with any degree of precision although my own experiments at the megalithic sites caused me to postulate another function for their construction related to the transmission of energy.

The purpose of exploring ancient energy in the modern world is to enable human beings to tap into this force for the first time in thousands of years. Such knowledge was believed lost to modern man because of the fragmentary nature of evidence concerning the ancient world. Giant pyramids, indecipherable glyphs, and other mysterious artifacts cover the globe but most records of the ancients have probably vanished, as a result of natural catastrophe, accident, or willful destruction by man himself. Thus, an intuitive recall of ancient times survives, along with some physical evidence, to help reconstruct the energy secrets of the past. Current investigations of the problem of ancient energy will take years of research, experimentation, and a reeducation about alternate methods used to acquire knowledge. Two separate kinds of methodologies in pursuit of scientific inquiry may be required by our institutions of higher learning depending on the nature of the phenomena studied. Measuring, or even speaking about ancient energy, is extremely difficult, given the limitations of present laboratory instruments and available vocabulary. With ancient energy we are involved in the investigation of a process and a product not yet addressed in conventional areas of study.

The source and the substance, therefore, of ancient energy depends on conditions in the outer environment, the interaction of human consciousness, and a belief system open to the concept that such energy exists. But the paradox is that the outer circumstances depend on the human dimension for the manifestation of the force and conversely. Since time and space play no role in explaining the force, cause and effect are reduced to one and the same thing. The process and the product become unified and the result is a holistic interpretation of the universe. The ancients may have understood these principles at a deep intuitive level. Some American Indian tribes still incorporate these ideas in their everyday living. The Navahos in the American Southwest build eight-sided hogans along with pyramidal structures, trying to unite positive and negative energies to increase the power at their sacred sites. In Malta, structures were built in the shape of the female body while in Egypt the pyramid form held sway. At every site, effective use of the energies appeared to depend on the shape of the structure, the human conductors, and the belief system of the culture which helped to sustain the force.

Ancient structures used simple geometry and mathematics to create the form. Frequently, the construction was circular, oval, or rectangular in shape, built in a dolmen style with post and lintel construction, open to the air. The dolmen style consists of massive standing pillars wedged into the ground as posts with a lintel beam across the top which rests on the stones. Other structures were pointed, domed, or flat, but most had inner chambers that were the locations of the most potent energies. For example, the construction of the oracle at Delphi, where the priestess conducted divination ceremonies in the inner chamber, may have influenced the incredible record of accurate predictions attributed by historians to the site. Sacred rites, held at shrines like Delphi and Ephesus, were often conducted in caves with domed ceilings which the ancients may have sensed were potent facilitators for the flow of energy. At the Hypogeum in Malta, used for divination, the dome is also found.

In addition, certain design elements were used to facilitate transmission of the force. Conductors were almost always encountered in the natural environment where early man observed color, form, shape, light, sound, and symbols. When I explored the ancient monuments of Europe and the Mediterranean, I began to form ideas about the conduction of ancient energy. I noted that certain design elements were used to enhance the force. Signs and symbols, like the single, double, or triple spiral were placed on most monuments along with swastikas, circles, and other universally recognized signs. I believe the form and placement of these symbols encouraged the energy producing potential of the monuments. In fact, the spiral form became a prime focus for my investigations, beginning with cave explorations in the Canary Islands. There, the Guanche Indians, now extinct, left their spiral imprints in the caves of La Palma. The same spiral, in a double and triple style, is found in Ireland at the caves of Newgrange where an enormous megalithic boulder guards the entrance to the cave with exquisitely carved triple spirals covering its entire surface. The same formations are also located on the temples of Malta and Crete and in other sites in Europe and the Mediterranean. After I observed the proliferation of the spiral in many archaeological sites, I began to hypothesize that it could represent the involution and evolution of the universe, or the yin and yang contraction principle believed to operate among all living things. In other words, the ancients may have known that energy is conducted most efficiently through a balance of the forces so they carved those symbols on the monuments to facilitate that balance. In nature we observe that the spiral is the basic form of the DNA molecule, the nautilus shell, and the intricate makeup of many flowers and plants. Following the observation of nature's principles, the ancients simply applied them to their own structures.

Light may also have been used as a conductor. Ancient monuments were built to allow just the right amount of light to enter at the appropriate time to create an energy effect. Examples of this can be found at the sacred temple of Dendera in Egypt and at Newgrange in Ireland, where I noted that a carefully arranged placement of boulders allows a beam of the sun's rays to fall inside the cave just once a year near Christmas time. The ancients were aware that planetary and solar energies were related to events on earth and that the alignments they created with their monuments served as giant magnifiers or actualizing forces that amplified cosmic activities. There is little doubt in my own mind that early man revered the action of the planets and may have known that management of energy forces on earth could affect the performance of heavenly bodies. Brad Steiger states in *Medicine Power* that rain dances of the American Indians have been shown to affect the action of weather in a very direct way. Myths connected with Atlantis or the Hopis indicate that ancient man believed he had the power to control actions emanating from outer space. In *We Are the Earthquake Generation,* Dr. Jeffrey Goodman discusses the accurate relationship between earthquake prediction, based on sensing the energies, and the action of the quakes themselves.

Crystals were also important conductors, many of which have been found imbedded in monuments and pyramids, similar to our use of crystals in radio and television sets. A crystal skull discovered by explorer F. A. Mitchell-Hedges in the 1920s in Honduras is believed to be a talking oracle by which ancients channeled and received energy, according to Richard Garvin, author of *The Crystal Skull.* Frank Dorland of Santa Barbara, lecturer at the University of California, reports that his work with the skull produced illumination from the base, causing distribution of images and light. He believes, like Garvin, that the eerie skull may have been an ancient sibyl used to aid in the transmission of ancient energy for divination. The Navahos are also reputed to use crystals in their healing ceremonies to help concentrate the beneficial energies toward the patient.

Another conductor used to channel the energies may have been sound. In fact, the vibrations of sound may have been the most important facilitator of all. The Egyptians used spells, incantations, and magic words, some of which can be found in *The Egyptian Book of the Dead*. Sound may have had the potential to alter physical form, heal, and induce motion. My own experiences and experiments inside the King's and Queen's chambers of the Great Pyramid in Egypt are dealt with in a later chapter. However, inside the pyramid's chambers, I observed that sound, at the correct frequency and speed, caused visible changes in matter. It also altered the consciousness of some of the participants in the investigations. The use of sound to accomplish superhuman feats can be noted in the Bible. When Joshua watched the walls of Jericho crumble at the sound of his trumpet, he may have understood how to use the forces of energy to break down barriers. According to the Biblical account, he asked the people of Jericho to remain silent for six days saying "Ye shall not shout, nor make any noise with your voice, neither shall any word proceed out of your mouth." On the seventh day, he told the people to shout, and the walls fell. In essence the withholding of the sound energy cleared the human conductors so that their vibrations had a heightened power at the time they were uttered. Certain words have been cited in myth and story to represent a sort of magic wand capable of inducing motion. The "Open Sesame" in *Aladdin and His Magic Lamp* may reflect an explanation of the unique power of sound to cause motion.

Finally, color may have been an ancient conduction device. All colors vibrate at different rates of speed. For example, blue, which represented spirituality to ancient man, is at a higher velocity on the color spectrum than either yellow or green. Purple supersedes blue on the scale and may have been the reason it was chosen as a royal color since its sacred vibration could have put the royal consciousness in touch with extrasensory energies. Colors were used on the faces of all ancient monuments including the Parthenon in Greece and the Treasury of Atreus at Mycenae. Since color is energy in motion, and varies its intensity according to the length and speed of the wave, ancient people ascribed sacred connotations to its varied qualities. It is my opinion that color, in conjunction with form and symbol, made certain early sites powerful conductors of ancient energy.

If you are experiencing a sense of remoteness about the phenomena of ancient energy, it is probably due to your lack of familiarity with the sites where the forces are most potent. Yet, in view of the explanations and examples given, I am sure you are already affected by it at this moment. The very process of inquiring into its nature will cause you to open your awareness to its possibilities. This is the first step toward making it work for you in your own life. Now it only remains for you to facilitate multiple encounters with the force, tapping into its reserves whenever you wish to draw the power of the ancients into present action.

We know that the world community has depleted existing energy reserves. Earth was given a warning of possible catastrophe years ago when scientists realized that our energy sources were vanishing at a frightening rate. Yet, the supplies that are threatened are basically physical in nature. We have not yet begun to experiment with those other energy sources, of which ancient energy may be a part. Perhaps, by investigating ancient energy, and integrating the results of that research with our knowledge of power drawn from modern technology, a new dimension would be available to modern man which would be startling in its possibilities.

For centuries ancient temples and tombs have lain dormant

across the globe, waiting for man to rediscover and activate their secrets. The awakening process will require great skill since it is based on a new and enlightened kind of investigation conducted by people who are willing to develop the consciousness and the belief system necessary to release the force. Those who undertake to tread upon this new field of energy research will bear a tremendous civic and spiritual responsibility. Yet, it is my opinion that everyone, at some stage of their moral and physical development, must be ready and willing to learn the principles necessary for the activation of ancient energy. For some people, interaction with the forces has already taken place. However, a deeper understanding of the nature of ancient energy will require time, patience, and a rethinking of present concepts about the nature of reality.

2. The Magnetic Mediterranean

The ancient energy of the Mediterranean altered my awareness and perception of archaeological sites. The impact of the energy, more than the places themselves, convinced me to begin my search for man's earliest beginnings in Spain where I had observed a direct connection between the quest for prehistoric origins and the principles of ancient energy. As I indicated earlier, the first members of the human race may have known about this unique force. They tried to express this knowledge through certain signs, symbols, and markings placed on the monuments. By becoming alert to these clues, my understanding and skill as an archaeological researcher improved. Formerly, I based my opinions of ancient cultures solely on the excavations of physical remains. As I learned to "think like the ancients," my view of history became more complete and I could interpret the signs as a means of understanding the way early people used design to facilitate the flow of energy.

In the Mediterranean setting I discovered an ability to communicate with the vibrations in the environment; as a result, I was able to make surprising observations, especially at caves and megalithic sites. Each time I reached an area where I knew, at an intuitive level, that discoveries could be made, I detected an increase in the power of the energy site. Apparently, I "picked up" the accelerated vibration and moved along with it until a physical manifestation occurred. In *Biofeedback*, Marvin Karlins and Lewis Andrews demonstrate a correlation between brainwave states and variations in the earth's electromagnetic field. During my research in Spain, I believe my own brainwave pattern responded rapidly to these force fields or energy centers.

My first interaction with the energies of the Mediterranean occurred at Nerja, where the guides believe a preflood culture existed. No written evidence supports such ideas, but my first investigations at the site led me to question the geological theory that the caves there were purely natural in origin. A long circular staircase led into the interconnecting caves which ran for blocks underground. Visitors observe irregular mazes of colorful forms covering the ceiling where newer structures were superimposed on the original shapes, obscuring any possible artifacts in the caverns. Since I did not like to explore the caves with a large group, I asked the guide for permission to investigate the area alone and walked toward the interior chamber where a large column with jagged edges ran from floor to ceiling. I could hear the noise of running water although I could not detect its source. Then suddenly a series of physical reactions led me to the evidence I was seeking; my heartbeat accelerated, my vision and hearing became more acute, and I experienced a general heightening of the sensory processes. Dozens of what appeared to be human artifacts, including art forms and sculptures, an ancient staircase, and a support for the second story of a building, became clearly visible in the dim light. All these objects were badly weathered, and several were attached by rock formations to another large eroded column. To the best of my knowledge, no one had made these observations before, even though trained archaeologists had examined the caves. I started to photograph everything in sight, but then chose to move as swiftly as possible through the caverns before the guide reappeared with the larger group. When he arrived, I told him of my discoveries challenging the traditional explanation about the natural origin of the caves. His initial skepticism about my report finally turned to curiosity and he asked me to point out the location of the artifacts.

Later, the guide mentioned that a speleological society, based

A powerful force field exists at the Caves of Nerja in southern Spain where manmade and natural artifacts have been deposited over thousands of years.

in Malaga, planned an exploration of the caves. I told him I was interested in their findings, so we exchanged addresses and became good friends through a long correspondence. Months later I received a letter saying that the Spanish group had reexamined the caves and concluded that the artifacts were of human design. The group assigned tentative dates to the cave of 20,000 B.C., aligning Nerja in chronology with Altamira and Santamamine in the north of Spain. The date also indicated that highly civilized people had inhabited the region prior to the traditional dates offered for the arrival of the Phoenicians at Cadiz (1600-1200 B.C.). In *Before Civilization*, Colin Renfrew demonstrates errors in our carbon dating methods, especially at megalithic sites and caves in Malta and Spain. However, even Renfrew's estimates are incorrect. It has been demonstrated that artifacts cannot be dated precisely by the Carbon 14 method since time appears not to be a constant and accelerates over periods of several thousand years. Renfrew's research caused me to consider the role of ancient energy in helping to establish general time periods for artifacts. If all of the best efforts to date prehistoric remains have been, at best, inaccurate, then human interaction with the ancient energies might result in a general chronological frame of reference for unclassified discoveries. My doctoral advisor, Dr. J. Manson Valentine, suggested this approach to me in attempting to date artifacts that did not respond to the Carbon 14 or other traditional dating systems.

My discovery at Nerja may be explained by Jung's theory of simultaneous occurrences which result from the existence of two different states of energy, the probable and the critical. My heightened perception from the powerful site induced the probable factor whereby I had the potential to utilize the data I was receiving by interpreting their meaning through my past knowledge and insights, stored in the cognitive area of my brain. Another person, lacking training in the field, might not go beyond the stage of the probable. Later, in the critical phase, I expanded my entire awareness by transferring communications from the probable stage across the perceptual pathways to produce the critical phase, resulting in observations and discoveries without any conscious reasoning on my part. This nonscholarly approach to the evaluation of information caused me to question the nature of scientific inquiry. Perhaps investigators were ignoring important nonmaterial data. For example, though I heard the sound of water, common enough in caves, the source was not visible. Yet it was a cue directing me toward the evidence I was seeking. Often I suspected the existence of human remains by "honing in" on just the right location. Jung suggests that a symbiosis, or affinity, operates among all things. I believe he refers to a common energy which pervades all elements of the universe and which ties us to all aspects of creation when we are in harmony with its rhythm. As I interacted with the force field and the artifacts, I felt "at one" with the environment.

Based on the experience at Nerja, I was more open to the reality of ancient energy at La Linea, a small village near Gibraltar. I had come to La Linea on a weekend break from my program of doctoral studies offered by the University of Granada in Malaga, specializing in advanced computerized linguistics. A friend, Antonia Ruiz, had invited me to visit her family, assuring me they had plenty of room. I soon discovered that the house was very small and had no plumbing or telephones. Yet the primitive accomodations were unimportant after I met eighty-year-old Antonio, the grandfather, a former seaman with a passion for Atlantis. On the second day, he suggested that I walk with him the twenty-four miles from La Linea to Gibraltar to view the archaeological sights. I was full of my usual vigor and imagined it would be no problem to take the hike, especially if an old man could walk the same distance. We travelled along the beaches where evidence of severe weathering showed on fossilized trees lying at the water's edge. Near the coast lay irregular layers of prehistoric ruins, mostly Phoenician, from the earliest recorded culture to have inhabited Iberia. Antonio believed older civilizations were buried in the substrata, since he had found several artifacts of unknown origin in La Linea. His memory, vigor, and endurance exceeded my own, and I was totally exhausted by the time we reached Gibraltar. We sat down on some boulders near the pathway that leads to the Rock and removed our last two oranges from the lunch sack. Before I had a chance to cut into mine, I spotted some irregular patterns on the paving stones beneath my feet. "Look, Antonio," I exclaimed, "look at these inscriptions." "Inscriptions?" he said. "What inscriptions?" "Antonio, the markings are inscriptions," I insisted, hoping he would agree. He ate his orange and continued to look puzzled. Without another word, I jumped up, camera in hand, and moved down the pathway shooting photos every six inches. I *knew* I was seeing inscriptions. Though I had never been to Gibraltar before, the markings looked strangely familiar. The chiselled look and the pattern made sense in concept, if not as actual letters. Antonio ran down the road after me, jabbering in Andalusian Spanish, convinced that the etchings were tool marks and not evidence of an ancient language. Ten minutes later, to his great consternation, I had taken over a hundred pictures. He demanded to know why I was wasting money on film and developing costs, since no scientific verification for my hunches existed. I noted in my diary that the scratches had been made by humans, and were possibly early Phoenician in origin. I remembered that Plato put the location of Atlantis near Gibraltar, and considered the possibility that the writing may have been left by survivors who fled the sinking continent. As soon as the photos were developed, I searched for a means to verify that the inscriptions were a form of ancient writing. Yet I was puzzled by my instant recognition of the marks. Such immediate "knowing" of information is an experience common to people engaged in the deciphering of unknown languages. Leon Pomerance, in *The Phaistos Disc*, states that a transitional stage occurs in the development of most forms of language when memorized symbols are recognized as messages before they come to represent individual sounds. He compares this stage of translation to understanding the larger meanings of road signs. The inscriptions I saw on the pathway at Gibraltar conveyed a message to me beyond the individual significance of the markings. I reasoned that perhaps the ancient energy emanating from such inscriptions may enter the human consciousness to produce a configuration which eventually results in a rational translation of the original symbols. John Chadwick in his book *The Decipherment of Linear B* reports that Michael Ventris, while deciphering Linear B Cretan writing, may have perceived a pattern to the language that revealed the underlying meaning. Thus, this might mean that another component, based on the principles of ancient energy, is at work in language decipherment in addition to phonetic translation. What the translator receives into his or her consciousness from the force fields which surround him gives meaning to the symbols.

At Gibraltar, there were other interesting facts about the ancient markings. The late Generalissimo Franco excavated paving stones from the caves above the area to build a walkway leading to the Rock. Dozens of stones were cut out and since the markings on them had not been previously recognized as writing, no search was made to find an accompanying bilingual text or entry key, necessary for the translation of a new language. Since I knew that the symbols could not be easily deciphered, my immediate interest turned to finding out why I had been drawn to the markings and if there was consistency to my discoveries that the inscriptions were actually a form of writing which had origins in the pre-Phoenician world. Subsequent correspondence with Dr. Cyrus Gordon, formerly of Brandeis University, indicated that the markings were inscriptions. But I was not able to pay anyone with the proper linguistic expertise to pursue my theory and I lacked the funds to continue with the translation. Often when I felt I was on the brink of some new discovery the data would suddenly stop coming in. My own interpretation of the causes for this perceptual blockage concerned the nature of the energies themselves. The stronger the field and the more heightened the vibration, the more evidence appeared. However, at certain stages in the process, the forces became too intense and the perceptual pathways temporarily closed down preventing me from assimilating the data. In other words, when I became overloaded with ancient energy, my consciousness "turned off" for a period of time. According to J. R. Oppenheimer quoted in *Time* magazine April 23, 1973, changes continually occur in the location and arrangement of protons, electrons, and other particles. These changes do not abide by traditional laws of time, space, or motion but are replicated according to acausal principles. My only hope of reopening the perception pathways was to repeat the experiences as often as possible.

The following weekend I continued the search in the Canary Islands where mysterious Guanche Indians (now extinct) were linked to the earliest civilized cultures in Africa. My investigations indicated that this tribe may have shared certain common physical and cultural characteristics with the Basques, including the same cranial capacity and a possible Rh negative blood factor. As suggested by Lewis Spence in *The History of Atlantis* this may have linked the two tribes to Atlantis. My interest in the tribe was sparked by their enormous height (they were reported to be ten to twelve feet tall) and their spiral petroglyphs which decorate the walls of caves in the outermost island of the Canary Chain, La Palma. These petroglyphs appear as etched markings in a spiral design, measuring about 6 inches in diameter, carved on the sides of the cavern used as a shelter by the Guanches.

My interest in the Guanche connection to both energy and Atlantis began in Tenerife at 9 p.m. on a Friday evening. After leaving the plane I went directly to the car rental desk. As the clerk explained directions to the city, a young man, Jose Magdaleno, offered to follow me to the hotel, claiming it was easy to get lost on the island at night. I was apprehensive about the idea, but the clerk assured me that Jose's uncle was an important town citizen and a fellow of the Historical Society. Jose waited while I registered at the hotel and then we went to his uncle's home. We entered the apartment and there I met Dr. Miguel Enrique de Carmona Sobrino sitting in a rocker reading a ten-pound volume of Don Quixote placed in his lap. Half an hour later he had planned my whole weekend program including a trip to La Palma to see the Gaunche caves and a visit with Dr. Luis Cuscoy, the local geologist. He also arranged for me to have a personal guide, Antonio Más Cordero. Dr. Carmona's hospitality was boundless and I couldn't believe my good luck in meeting Jose and his family.

The next morning I started out with Antonio to visit Mount Teide. The ride was long and arduous, but my fatigue vanished when we reached the mountain. Antonio warned me that the ascent would be difficult, but something in me impelled me to reach the top; despite his admonitions, I kept climbing even though it took an hour to reach the peak. Fortunately, Antonio stayed right behind me. My hunches were not in vain. Two feet from the summit, I spotted a reddish brown shiny area, which according to the guide was a segment of a fossilized cedar tree. I wondered what it was doing at top of Teide. It looked exactly like the fossilized cedars I had observed on the beach near Gibraltar on my long walk with the older Antonio. We discussed the specimen at length and Antonio told me that a great upheaval may have pushed the tree to the top of Teide. The next morning, Dr. Cuscoy explained why this happened. He said that the islands were the peaks of submerged mountains in a great Atlantic Ocean land mass that had been drowned when the seas rose following a great catastrophe. The Canaries sloped all the way down to the bottom of the ocean floor without the usual underwater platform extending out from most islands. The remains of the Guanches were found on the mountaintops. Due to their great size, it is possible that they could thus have quickly scaled the mountain at the time of the flood in order to survive the holocaust. The possibility was fascinating. I could have spent all day discussing this controversial subject, but my afternoon was heavily programmed.

After an hour-long flight from Tenerife, Antonio and I landed on the tiny runway at La Palma. Isolated in the Atlantic, this timeless island is dotted with craters, jagged rocks, and lava formations. My first impression was that it looked like a scene from a NASA moon shot. Yet I was caught up immediately in its strong energy vortex and had the urge to go walking through the black volcanic dust that covered the island.

The limited time, however, demanded that I concentrate on the Guanche caves in order to carefully examine their double and triple spirals. Dr. Carmona had a personal driver waiting for me at the airport and he explained the geology of the island enroute to the caves. After we arrived, I crawled in and out of the crevices, touching the spirals and taking photos from every different angle. The symbols conveyed a definite message, I was sure of that. Yet I couldn't discern their meaning, probably because I hadn't been in the caves long enough to absorb the vibration. I knew that the tribe was known for its enormous vitality; perhaps that strength still permeated the caves.

I suspected that the spirals were not an accidental motif. Jill Purce, in *The Mystic Spiral*, infers that such forms convey a message about the nature of matter which moved in curves and spirals, colliding, recoiling, and exploding. I believe the Guanches may have tried to represent moving masses of energy when they carved their symbols into the rocky edges of the caves just as the Hopi represented Mother Earth with a spiral, signifying creation or regeneration of energy. In ancient Mediterranean tradition double spirals meant spiritual growth, and triple spirals described the separate essences of the universe—physical, metaphysical, and etheric. The single spiral was a universal symbol, but the double and triple versions were found abundantly in Europe, the Mediterranean, and the American Southwest. Ancient people may intuitively have understood the relationship of consciousness to the balance of subtle energy currents. Plato interpreted the significance of the spiral as the involution and evolution of the cosmos, or the contraction and expansion of the universe. It is interesting to note that the DNA molecule, the nautilus shell, and other natural objects all follow the same pattern.

The possible connection between spirals and the flow of energy absorbed my interest. Each time I was in an area where spiral petroglyphs were etched into megalithic boulders, walls, or caves, I experienced a more intense sensation from the field of force, which heightened as I moved closer to the symbolic forms. Spiralled sites also had longer residual effects than other areas, causing me to feel the energy flow long after exposure to the sites. Sometimes the effect endured for several weeks. After noting that spirals seemed to trigger the energy centers into action, I sometimes became discouraged at other sites where no environmental clue activated the process. Yet I was comforted in knowing that my receptivity always opened if I allowed sufficient time to pass. I knew that ancient energy was elusive, but I believed that as in all research, there had to be some consistency of experience. I left the Canary Islands, still absorbed in this problem.

My next stop was Bilbao, capital of Spain's "Campo de los Vascos," or Basque Country. Antonio Cordero had given me a ring inscribed "Long live Arantzazu" in Basque. Enroute to

Bilbao, fellow travellers explained that Arantzazu was a convent in the mountains of northern Spain where, according to folk legend, someone had seen a tiny virgin standing on a tree branch in front of the caves, located behind the chapel. My eyes lit up when I heard the word "caves," but the convent was far from the main road to Bilbao and I decided to explore it at another time. Apparently, visions like the one at Arantzazu were frequently reported in Spain by so-called "normal" people, an idea I rejected even though skeptical friends occasionally called me Santa Teresa when I discussed my theories about energy. As I considered the story of Arantzazu, it seemed plausible that the force field near convents and churches might cause some people to envision or bring forth images. Henry Sorge, a dowser, once told me that his rod went out of control when he tested churches in Mexico, traditionally areas of high energy. Ancient and medieval chapels were built on top of pagan sites known to be powerful energy centers. The vibrations around the convent may have made images materialize for some people, giving rise to stories of visions; such stories were common in Mexico City near the main Cathedral of Guadalupe.

I made note of the Arantzazu caves in my diary, assuming that a larger framework existed into which the bits and pieces of information would eventually fit. That framework could only be provided by the rules of quantum mechanics. In approaching matter through this theory, reality as we perceive it may be an illusion. Like a horse fitted with blinders, we see only part of the essence. In searching for a larger reality to encompass information from the material and nonmaterial data I had received, the puzzle only grew larger. There were imponderable gaps in the evidence, and my frustration continued to increase. Yet I suspeced that my explorations into the lands of the Basques might corroborate the experiences I had encountered in the south of Spain.

Bilbao was a stopover point for my main destination, the caves of Santamamine, buried deep in the heart of the Basque farmland. I drove for 20 miles in the mud and rain, arriving within half a mile of the caves when my car motor quit and I was stuck on a dirt highway. By the time the auto was repaired, it was four o'clock but I drove on, figuring I had at least an hour to explore. There were no signs of tourists parked along the highway so I left my car at the foot of the incline and climbed up an endless series of stone steps leading to the entrance to the caves. There I was surprised to see several people with a guide ready to begin explorations. For a change, I was comfortable in the midst of a group.

The guide pushed the iron door inside and we moved slowly into the interior, walking about a dozen yards toward the center where beautiful paintings and art work lined the walls. By the time we reached the middle of the tunnel, claustrophobia and/or heavy energies in the cave told me to return to the entrance. The heavy vibrations I experienced in the north of Spain contrasted sharply to the buoyancy I felt in the south. I hadn't been in Bilbao long enough to make the adjustment and was deeply affected by the force field of Santamamine. I told the guide I was sick and would have to make my way back alone. He felt my move was ill advised because the path back was dangerous and deceiving. I assured him I was a veteran explorer and knew the way. Ten minutes later, I was totally lost and very frightened. The caves were well lit, but the lights were scheduled to go out at any time. I reached for my flashlight, flopping down on a flat rock to look over the situation. As I probed with the light, I noticed vivid colors on the walls, mostly greens, yellows, and reddish browns. In the center was a petroglyph that appeared to be a fern. This area had had a definite purpose in ancient times, possibly as a divining chamber or ritual site for primitive ceremonies. An altar stood in the corner and on the floor were various artifacts which looked as though they may have had religious significance. I ran my hand over the fern-shaped glyph, an unusual design for cave dwellers, who generally drew animals as their prevailing motif. I knew from my studies that Cro-magnon man had once inhabited the caves but I also remembered that the origins of this race had never been ascertained. Cro-magnon's roots are so obscure that anthropologists can only guess why this race was more highly civilized than its contemporary, Neanderthal.

My interest, however, centered on the fern, which I sensed

had definite linguistic significance. It was a symbol, used by the ancients to record and store information the way we presently use computers to enlarge our memory capacity. I believe the importance of the glyph lay in a more encompassing meaning and not in its exact phonetic or literal interpretation. The fern turns up repeatedly in caves throughout Britain and Ireland. Later, I saw it on a chart of early Phoenician symbols, and finally it appeared on pyramid building bricks at Comalcalco, Tobasco, Mexico. In Indian writing, certain symbols were universally understood. LaVan Martineau, in *The Rocks Begin To Speak*, states that major ideas, projected by symbols, were interpreted in the same way, regardless of culture or geographic location. This led me to believe that ancient writing could be reduced to a small number of glyphs, widely understood to refer to universal concepts.

The energy vortex of Santamamine was as powerful as those near the Etruscan tombs and on the Canary Islands. Spain has many of these ancient force fields which alter perception and increase the flow of timeless energy. These energy centers have a powerful effect on visitors to the Iberian Peninsula. Dr. Gerson Cohen, commenting in the NBC television special, *Sunlight and Shadow: The Golden Age of Spanish Jewry*, stated that ancient Andalusia was "gone . . . yet it was there." He referred to the powerful mass of energy near Granada, Cordoba, and Seville, which I also experienced. World-famous Spanish literary critic Emilio Orozco Diaz, told me that "todo es posible en Granada," everything is possible in Granada. He described many miracles that had occurred in the city. Gypsies who lived in nearby caves cultivated and reinforced the vibrations, making Granada the most powerful force field in Andalusia. The legacy of prehistory is ancient energy. It has not changed in Iberia for thousands of years.

The caves of Santamamine may have been used for sacred ritual, which reinforced intense vibrations natural to the site. Inside, a fundamental unity induced a feeling of peace, almost like a hypnotic state. Within the center of the cave, there was a basic harmony which caused me to feel a kind of euphoria. (Many of the ecstasy states of Eastern mystics were induced by meditating in high energy centers like Santamamine.) My initial fears of becoming lost were assuaged and I stayed inside the cavern after closing time when the exit gate was sealed. I beat on the door with my fists, attracting the attention of the guard who lived below the site. He opened the lock and I drove hastily back to Bilbao. In my exhausted condition, the modest hotel seemed like the New York Hilton that night.

I flew back to Malaga the next morning and left immediately to explore the megalithic complex at Antequera, high in the mountains of Andalusia, near Granada. The caverns were were built in dolmen style with post-and-lintel construction. Ancient people constructed monuments in strategic locations across Iberia, the British Isles, Wales, and the Orkney Islands to keep energy in motion. The dolmens were placed in this way to set off a chain reaction of energy that was focussed at each individual site and collectively at all the sites. I believe dolmens were power points for the conduction of ancient energy. Each dolmen post was its own energy field, receiving vibrations from other megalithic sites. The energies picked up momentum from one power point to the next. Although the initial reception of the energy at each site was random and acausal, the total effect reached a powerful crescendo at certain points along the line. Contemporary man experiences such reactions randomly, but ancient man planned spatial relationships in order to increase the energy potential. In *The Standing Stones of the Lothians*, Adam Mclean points out that pillars were erected at points on the earth where telluric currents took form. These were power centers or energy points, known intuitively to the builders of the megaliths.

Antequera's dolmen-style caverns were constructed in the mountains rather than on flat land or underground. A local guide explained the discovery of the caves and told me that the early inhabitants had left no writing. After his speech, I walked into the corridor of the first cavern where bare walls were supported by one tall standing stone. What appeared to be symbols similar to some I had seen some years earlier in the basement of the palace of Sintra in Portugal were carved into the pillar. The cave was warm despite the inclement weather outside; in fact it exuded an almost cozy feeling. I sat down and studied the symbols, drawing them in my notebook with the usual subjective commentary. The guide admitted that it was the first time he had seen them. Archaeologists had overlooked the markings or had not realized that they were a form of writing. Some ancient people wrote messages that had no regular word-order pattern when translated consciously by the receiver. To scholars who searched for prehistoric writing with a pattern, the random scratchings were not meaningful. Our own system of written communication includes sound, form, and a regularity to contextual clues within the structure of the message. The ancients, on the other hand, may have provided a framework for their concepts by applying energy channeled through the mind. I often identified writing, as well as artifacts, in this manner. For example, I was able to see new meanings at archaeological sites such as Stonehenge where I photographed a face on a large standing stone that had gone unrecorded in thousands of published pictures of the site. My recognition of the markings on the pillar may have occurred because the pillar transmitted an energy to me which was not picked up in consciousness by other researchers, much like the unfamiliarity one experiences when confronted with a foreign language for the first time. Other researchers may also have noted the markings but, lacking knowledge of the configuration, dismissed the face as a valid find for lack of conscious reinforcement from previously observed markings of a similar nature. Recognition of archaeological finds may be inconstant among researchers depending on the level of their awareness.

The "now you see it, now you don't" phenomena among scholars may possibly be explained by Einstein's theory, since the physical appearance of an artifact is altered by the investigator's perception of it. Physical objects are believed to take on form because they vibrate at lower rates than nonphysical entities. For example, we can view the sun as a physical form but we cannot always perceive the fine gases that escape from it because they accelerate their rate of vibration and move out into the atmosphere, beyond the range of our sensory processes. Perhaps a finely tuned searcher may be able to perceive objects on the visual plane that go unnoticed by others. In *The Gold of the Gods*, Eric Van Daniken claims he saw gold plates and objects in

Megalithic remains at Stonehenge England may transmit energy from one standing pillar to another, acting as power points in a chain which extends to sites in Europe and the Mediterranean.

South America though other scientists never were able to verify his discoveries. If some researchers perceive ancient ruins in different ways, then perhaps our experiences affect the reliability factor in discovery. Thus monuments, such as the pyramids, endure because mass consciousness in connection with their form, has remained unchanged throughout the centuries. The energy force which motivated the builders of the pyramids may have been so powerful that it stayed in the unconscious awareness of many civilized cultures in our history. In present times the pyramids continue to exist because their massive structure is reinforced by the consciousness of millions of observers who are drawn to the energy centers. This possibility, though tentative, led me to believe that ancient energy might be a tool in archaeological discovery.

The next investigations after Antequera took place in Malta, fifty-eight miles south of Sicily, where evidence of advanced prehistoric cultures exists. Twenty-eight miles of underground tunnels interlace the island of Malta with connecting passageways, rooms, and dolmen chambers used by ancients for secret ceremonies. At ground level, there are half a dozen major megalithic temple complexes and dozens of smaller ones in the style of Spain. Some sites have representations of phallic symbols and mother goddesses. One temple, Hagar Qim, is formed in the shape of a woman's body as seen from the air. In Malta I sensed a connection between the phenomena of ancient energy and the sexual and creative forces. Energy, to be effective, must be released. Sexuality was one energy-releasing agent practiced in the ancient world. Sacred ceremonies at Delphi, Ephesus, and Chichicastenango in Guatemala used sexuality in divination and prediction, since it cleared consciousness and facilitated the flow of vibrations. At one Maltese temple, I saw half a dozen male and female phallic symbols indicating the importance of these sexual rites.

Ancient energy was powerful in the underground caves and chambers of Malta. One cavern, the Hypogeum, at Hal Saflieni was formed in the dolmen-style and contained an ancient oracle, reputedly activated by the sound of a human voice in early times. A projecting ledge in the oracle's room was hewn in the rock which according to A. J. Aguis, might have been specially made to carry the sound and provide peculiar sonorous arrangements. A human voice, vibrating at varying frequencies, was used by the ancients to direct forces and to secure information, unavailable through any other source. Energy was evoked by incantations chanted at altars near the site. To receive a message, the diviner projected his voice through a round hole cut into the oracle room. One or more questions were asked, and answers were received from vibrations that reverberated through the opening. The Hypogeum was probably one way that early people obtained knowledge.

Documentation exists concerning such divination. In *A History of the Delphic Oracle*, H. W. Parke describes how ancient Greeks received accurate prebattle advice at the oracle during the Persian Wars. The Egyptians are believed to have used the inner chambers of the Great Pyramid for the same purpose while the oracles at Ephesus and Dodona provided the same outlet for divination.

In the Hypogeum I felt rejuvenated and tried to activate the oracle, though I was told that only a male voice could receive a message. The guide explained that the ancients used the divining room to send and receive messages. I practiced transmitting

The author's daughter, Laurie, peaks out from the interior of an ancient construction in the Hypogeum, a high energy center underground on the island of Malta.

sound into the opening for thirty minutes. Finally I heard a noise, definitely not an echo, different in pitch and frequency from what I had transmitted. In searching for a possible explanation for the returning vibration the ideas of linguist Edward Sapir made sense. In *Language*, Sapir infers that languages have an inner sound system, unconsciously known to the speaker. The returning pitch from the oracle may have been an answer to the inner message of my voice, which returned as a symbolic representation, to be changed into concepts at a perceptual level. As I sent forth the vibration into the opening, I also may have emitted another energy, different from the wavelength created by the sound of my voice, which interacted with the same frequency present in the oracle. I knew the sounds were a form of information, but I could not understand their meaning. My experience was comparable to that of animals, who sense general ideas through levels of awareness not presently developed in humans. Pets have been known to predict illness or death in their masters, possibly by detecting a change in the rate of vibration to which they are normally accustomed. Bill Schul, in *The Psychic Power of Animals*, states that many creatures are more aware than humans of subtle energy fields and vibrations. We can learn from the sensitivity of animals to predict earthquakes and other cosmic events. The reception of knowledge is the same in humans as in animals. The difference is that finely tuned sensing devices remain open in animals while in man, reception is usually blocked by intellectual processes.

The energy field of Malta may have led to highly developed techniques of construction at the sacred sites. I believe early Maltese used symbols to transmit vibrations at the temples just as modern architects use lines and curves to create the appearance of flow. On Malta, spirals and random dots carved by the ancients performed that function. Such designs appear in sites across Europe and the Mediterranean and are not, I believe, coincidental. Wherever these markings were found, I experienced maximum energy flow. The ancients may have concentrated on the symbols in order to produce an altered state of consciousness which allowed for movement and flow. Even when ceremonies were not purposefully undertaken, the symbols kept the energy mass in motion in and around the sacred areas.

In search of more symbols, I flew to Catania, Sicily, where the positive effects of the high Maltese vibration immediately left me. Several years earlier, before I had become interested in the phenomena of ancient energy, I had explored neighboring Taormina. My memory of the positive environment there encouraged me to return to other Sicilian sites, but I was disappointed at the outset by the dense force field of Catania. My daughter Laurie, who worked with me in Malta, agreed that the vibrations were almost impossible to bear. To us, Sicily was a reminder of warfare and death. Since there were no megalithic temples or caves there, and nothing of real importance for my work, we left Catania and headed down the eastern coast toward Syracuse. Once on the road, we felt better immediately, but I was still in shock from the rapid shift in vibration. This experience alerted me to the danger of exposing oneself to the fields of force without proper protection. One's ability to channel and transmute the essence is critical in the investigation of ancient energy. Catania was definitely not a good place for the inexperienced researcher to begin, probably because Sicily had been a battleground for centuries. The negative vibrations still remain in the environment, perhaps reinforced unconsciously by the present inhabitants.

At any rate, we were surprised to be rejuvenated as we approached Syracuse. Five minutes after we entered the city, we saw a large area of caves on the cliffs which were still under excavation. Layer after layer of rock revealed the remains of prehistoric cultures. I explained to Laurie that Sicily had served as a natural survival site for refugees fleeing disaster in the ancient Mediterranean world. It had been the melting pot in antiquity.

The caverns at Syracuse emitted the same warm feeling as the caves of Santamamine and the underground caves of Malta, where the intensity of vibration increases as one walks toward the interior chambers. At Syracuse, the caves were shallow, which diluted my ability to sense information. I still was not in tune with the force field of Sicily because I lacked the ability and desire to interact with energies I felt were incompatible with my own. I decided to abandon my research there and begin investigations in Greece. The next morning my plane landed on Crete, in the middle of the Aegean Sea.

In 1903, Sir Arthur Evans excavated the palace of King Minos at Cnossos, basing his search on the myth of Theseus and the Minotaur. The discovery, made shortly after excavations at Troy by Heinrich Schleimann in the nineteenth century, opened up a whole new field of archaeological exploration based on the use of myth and intuition. According to their biographers, both Schleimann and Evans used intuitive data to locate the sites as well as information from myths and legends. They knew exactly where to excavate and could reconstruct cultures based on traditional and nontraditional sources of information. Other stories echoed reports about Henri Layard in Assyria, who uncovered the remains of Nimrud after having a clairvoyant dream while sleeping near the palace. I suspected that in these discoveries and in others throughout history, ancient energy played an important role.

The greatest concentration of energy at Cnossos appeared to be underground in the spiral labyrinth leading to the center of the palace, precisely the location where each season the legendary Minotaur devoured certain Greek children who were selected for sacrifice. It occurred to me that perhaps there had not been a Minotaur at all, but rather a high concentration of energies at the center of the labyrinth which overwhelmed the youths. The Minotaur may have been a symbol for the untamed power of the forces.

After exploring the palace at Cnossos, I examined samples of ancient Cretan writing, much of which still puzzles linguists. Cryptologist Michael Ventris had broken the code of Linear B writing, but Linear A was still untranslated. The meaning of the symbols on the Phaistos Disc, an inscribed plate discovered at the western end of Crete, was also unknown. It occurred to me that perhaps the Cretans may have used one system of symbols to transmit energy and another for written communication. This could explain why languages like Etruscan are untranslatable even though the symbols within them can be identified. Perhaps scholars failed to decode messages that also functioned as energy transmitters, since they didn't perceive the dual function of symbology in the ancient world. People of prehistory availed themselves of information from other dimensions besides the strictly causal plane. Thus, this may be why our present records of history are linear and without holistic perspective.

The Cretan force fields were as potent as those in Malta and Spain, but I abandoned my work in Europe in order to examine areas of my own country. The signs and symbols of the Mediterranean led to my first tangible experiences with ancient energy and its possibilities for use in the decipherment of languages. I realized that energy could be heightened through the strategic placement of monuments and forms, especially at megalithic and dolmen-style sites. I learned I could be "drawn" to artifacts and information by simply connecting with the power of energy centers used by early people. Most of all, I gained supporting evidence for the many random occurrences I thought were accidental in my early research. After Europe my next investigations took place in the American Southwest where I observed that ancient energy had other important functions. There, my increased fervor led not only to archaeological discovery but to new knowledge about techniques of healing, control of the weather, and information about the ways ancient people controlled the forces in their environment. I had been unaware of ancient force fields in California, and was surprised to discover that locations in the American Southwest produced a reenactment of my energy experiences in the Mediterranean.

3. Ancient Energy and the American Southwest

The ancient energy of Arizona definitely affected archaeological discovery. Fields of force at Hohokam Indian burial grounds near Cashan led researchers to locate artifacts using nontraditional approaches. For example, Tony Arlotta, a lay archaeologist, instead of "trenching" or probing the soil, found burials from clues received intuitively at the site. He also dreamed about pottery that was excavated according to his description of the area the following day. I discussed Tony's abilities with Tucson archaeologist Jeffrey Goodman, who, in his book *Psychic Archaeology*, described similar experiences at Flagstaff, where psychic Aron Abrahamson predicted the site at which the earliest relics in North America were later found. Tony and Aron had much in common. They literally "saw" through the ground without the use of conscious reason or any physical evidence to clue them in to results. I asked Tony if he would test his talents in front of newsmen and scientists. He agreed without hesitation, so I secured a new objective site for the experiment, donated by the Pima Maricopa Indians.

Our small group gathered at 10 A.M. and waited for the newsmen to arrive. Tony's discoveries began with a twenty minute walk across the site. Stopping abruptly as if listening to a sound, he probed around for a few seconds and then went into action. The burial of human remains and artifacts was excavated intact. Tony smirked, "I always come through, don't I?" he said. The television reporters were overjoyed. They photographed the two of us sitting in the trench, and compared his ability to Heinrich Schleimann at Troy. Schleimann, like Tony, said "Dig there," at Hissarlik in Turkey. Under nine layers of ancient cultures lay Troy.

We were big news on Phoenix television that night. Even the reporters demonstrated their use of dowsing rods to make smaller discoveries on the site that day. The art of dowsing rests on principles of ancient energy. Verne Cameron was a skillful dowser

Intuitive archaeologist Tony Arlotta examines a Hohokam Indian urn which he excavated in Arizona by "seeing" into the ground where the ruins were located.

who discovered water at a dry bed, now Lake Elsinore in California. In his book *Map Dowsing*, Cameron states that a dowser's accuracy results from three energies: the physical, the metaphysical, and the etheric. The ancients also identified these distinctions. They saw the three components as one unified whole which was channeled through the human body during the act of dowsing in harmony with the natural balance of the universe. Frank Brown of Northwestern University theorizes a connection between dowsing and the earth's magnetic field, suggesting that a space-time grid exists for all living creatures and that the dowser responds to such cosmic cues which direct him to what he is seeking. Brown also points out that bioelectric activity of the brain is related to the force field of the earth, resulting in the ability to detect or sense information not immediately perceptible in the physical plane.

I was less interested in finding artifacts than in Tony's approach to archaeology. I watched him as he detected a burial mound by walking around the site. He instinctively knew in which direction to go and where to begin digging. I suggested that he might be communicating with the ancient energy flow of the tribe, tuning into a fine vibrational essence that other investigators did not perceive. In *Activation of Energy*, Pierre Teilhard de Chardin points out that an energy of unification exists, which emerges in the human body assisted by physicochemical forces. Relating this theory to ancient energy, I theorized that de Chardin may have been referring to the integration of the physical, metaphysical, and spiritual essences which require human consciousness and a belief system to bring it into existence. Tony accepted such ideas, since he and his wife Sherry, without extensive textbook knowledge, had experienced what seemed to be a collective memory of Hohokam culture. Tony's educational background was average but his religious training was strong. This element of faith, coupled with a powerful intuitive sense, gave him the confidence and ability to make substantial finds.

I decided to try my own luck at the burial grounds to see if I could perform in the same relaxed manner as Tony. Unfortunately, I was a dismal failure. Even though I submerged myself in the vibration of the site, I could not find anything other than potsherds. The more I consciously tried to make a discovery, the less successful I was. I suddenly realized that Tony facilitated the action of the energies while I tried to direct them. Tony blocked out other people's vibrations; I couldn't even screen out my own. Frustration increased as I watched my daughter Laurie excavate an Indian grinding stone that she "knew" existed before lifting one shovel of dirt. Laurie's abilities were like Tony's. I had the feeling that my strong academic background was obstructing the flow of energies when I consciously tried to make them perform. My Mediterranean experiments had been spontaneous, and therefore successful. The Arizona experience failed because I was trying too hard to control the ancient forces. Energy freely converts from one form to another and flows in an uninterrupted channel unless it is blocked by human consciousness. I had to train myself to conduct the vibrations rather than forcing or inhibiting them.

The energies of Phoenix had restorative and uplifting effects. Layers of Indian cultures were buried under skyscrapers and apartment complexes, producing force fields of great intensity. Many holistic healers live in Phoenix who induce cures by redirecting the natural flow of vibrations into their patients' bodies, transferring energy from the power source available to them in Arizona. When I first observed their treatments, I discovered that they did not follow a particular method, but received the energies spontaneously as I had done in the Mediterranean. Further research indicated that the direction and use of vibrations was not always an accidental process. In *Breakthrough to Creativity*, Dr. Shafica Karagulla describes medical doctors who diagnose patients by "seeing through them." She documents cases of physicians who gave examinations without radiography by perceiving damaged energy centers in the body. Healers would clear their minds through meditation and other energy-releasing techniques, creating a clear channel for the flow of energy. The book *TM* by Harold Bloomfield, et al., demonstrates that meditation produces a harmonious coherence of the brain's electrical interaction between theta and beta waves, corresponding to an expanded awareness. This may indicate that transcendental meditation can potentially clear the brain's perceptual pathways and allow the higher energies to flow into consciousness where they help synchronize brain waves. Once the consciousness of the healer is adjusted by this means to the energies, a higher vibration can be conducted to the receptive patient who receives the healing flow.

The Association for Research and Enlightenment operates a medical clinic in Phoenix. Working with holistic healing methods suggested by the late Edgar Cayce, the ARE clinic treats thousands of people yearly by rechanneling body energy. They also give lessons on proper spiritual and dietary habits, following Cayce's philosophy that the cells within the body eventually become one with the creative force we call God. Healing energies, conducted through the mind, body, and spirit of man, are one.

Cayce drew many devoted followers, since he was one of the first leaders to discuss nontraditional ways of knowing. But I had not heard of Cayce in the academic world, nor had I considered using the principles of ancient energy in my own field. As the movement to raise consciousness assumed global proportions around 1965, I made no systematic effort to change my own awareness. A more difficult problem was to unlearn preconceived ideas about the nature of scientific evidence. At the university, professors never mentioned nonmaterial data. After I decided to investigate the elusive new phenomenon, there was no documentation, scientific vocabulary, or instrumentation to rely on in connection with my work. At times I felt despondent over the perplexing nature of the problem, since I always lost the guiding principle just when the data started to make sense. However, my ideas began to crystalize in Phoenix, where I met a sensitive, educated woman named Moira Carney who worked with Navaho students at the community college in Tsaile. Our discussions confirmed my own intuitions about ancient energy and were reinforced by a week of investigations conducted on the campus at Moira's invitation. I arrived at Gallup, New Mexico, during the Intertribal Festival, where Moira and two young Navaho friends were waiting to drive me to Tsaile. On the way we discussed the significance of Indian oral history, especially stories about the land between four sacred mountains where the tribe had migrated after a great flood. The college was built within a larger region known as the Navaho Nation, deriving its name from the province of Navarra in Spain. Mitch Uribe, a specialist in Navaho tradition and ancient energy, helped me understand the sacred wisdom of the tribe. Half Basque and half Navaho, Mitch had a master's degree from a major university, but his real interests centered on Navaho oral history and the healing practices of the medicine men. After graduation, he had returned to work at Tsaile, but the administration insisted that he teach traditional academic subjects on the campus. Most Navaho leaders preferred to modernize the curriculum. Mitch was an exception.

The morning after my arrival, he suggested that I walk through the campus to observe two distinct types of structures. One was a traditional eight-sided hogan used for student housing, and the other, a larger building, was almost pyramidal in shape. The pyramid served as a center for learning and administration. Mitch explained that the eight-sided hogan represented the female, or negative, energies while the pyramid channeled the male, or positive, energies. Together they created a polarized vibration on the campus, drawing their strength from the sacred land. Mitch claimed that medicine men, working in a room one-third of the way up the pyramidal structure, healed patients by encapsulating their bodies in a flow of positive energy. He indicated that healing was achieved using incantations, color, light, and sound. Modern experiments have demonstrated that light and color can alter the flow of energy in the body. Japanese researchers at Kyoto University report that the neuroendocrine system is color sensitive and reacts well to a variety of auditory stimuli. Variations in light can change adrenalcortical hormones. When I heard about Navaho medicine men, I surmised that a principle similar to the Japanese experiments took place as the medicine men actually appeared to reverse the nature of the body energies into a state of well-being.

I had read about Navaho medicine before coming to Tsaile, always questioning the authenticity of such mystical approaches to healing. But I observed the Phoenix practioners who, like the medicine men, actually cured patients. Other Indian tribes also had healing techniques. In *Warriors of the Rainbow*, William Willoya and Vinson Brown describe cures among the Sioux tribe. For example, a young boy with pneumonia was given physical remedies but was finally healed by means of the principles of energy and spirit.

Indian tribes like the Navahos also directed energy forces toward the control of weather and other natural events. The production of rainfall through Indian approaches was tested by archaeologist William Payne. In *Medicine Power*, Brad Steiger reports that Payne induced rainfall in twelve out of thirteen attempts in clear weather, using effigies and practices of the ancient Zapotec Indians.

At Tsaile, I learned bits and pieces about such ideas, but the students I interviewed were reluctant to share anything of great importance. Mitch was more cooperative. On the evening of the second day, I was resting on a Navaho blanket at his apartment with Moira and Mitch's girlfriend Debbie. We were discussing Navaho conceptions of time, space, and the universe when a small spider crawled across my leg. I reached down and smashed it. Mitch leaped off his chair horrified.

"What have you done?" he said. "We don't kill spiders here."

"You don't, why not?" I asked.

"The Navahos do not disturb the flow of energy that moves freely among all living things."

I was flabbergasted. "Well then," I demanded, "how do you harness nature? Your college would be overrun with insects, rodents, and all manner of things if you didn't have some form of control."

"We just send out the proper vibrations," Mitch said quietly. He went on to explain that an entity can be altered by directing a flow of energy toward the object. He described the experience of a Navaho construction team building a hogan close to a mountain. After they had cut into the hillside, an enormous spider came slowly out of the underbrush. Instead of killing it, the workers began to chant in order to change the energy flow. Slowly the creature moved away and the work proceeded.

Mitch explained that the elders of the tribe knew the secrets of making natural forces work for them but withheld this knowledge from any other than tribal members since people who were not in harmony with Navaho energy might dilute their own powers as well as the collective strength of the group.

Eventually Mitch and I gained more confidence in one another. Although he shared many Navaho secrets, he made it clear he had to keep certain information from me. He innately distrusted strangers, claiming that Navaho oral history had been secretly removed to Mexico and the Caribbean, possibly to prevent its widespread dissemination to groups outside the Navaho Nation and to protect the information from destruction through natural catastrophe or human intervention. He suggested that we collaborate on a project to send the medicine men from the Tsaile to collect the missing information, claiming that the oral traditions contained references to the use of energy by the ancients, and wisdom which would revolutionize our modern conceptions of time and space. Mitch's views, unorthodox for a contemporary Navaho, were based on his knowledge of ancient force fields, part of the Navaho heritage known only to select members of the tribe.

I witnessed another form of communication at Tsaile. Navaho students often practiced telepathy, engaging in eye to eye contact for long periods of time. In my travels I observed the same phenomenon among the Mayans of Guatemala and the Cuña Indians of Panama who use this ancient way of interacting, accompanied by rapid eye movements and short, terse speech. Teilhard de Chardin theorized that one day we would construct instruments capable of recording rays emitted by thinking brains and channeling them in a given direction. We also may be able to decode unspoken messages that carry powerful vibrations, and are part of the dynamic social structure of certain ethnic groups. Exchange of ideas through nonverbal communication is a transfer of ancient energy, useful in maintaining patterns of culture. Tribal members receptive to the powerful force fields of Tsaile tuned in to the ancient wisdom. Generally it was a small group of younger Navahos who opened their consciousness to the higher vibrations, while the elders preferred to follow the path of technology in order to keep pace with the swiftly moving Anglo culture. While growing numbers of young people maintained cultural traditions, there was a division among the elders of the Navaho Nation. This schism probably came about as a result of the changing awareness of the youth, who rejected the elder's submission to Anglo standards and demanded a return to their Indian heritage.

The energies at Tsaile did not directly influence my consciousness as they had in the Mediterranean, where I had been drawn into the force field without any rational effort to channel or direct the vibrations. Things had just "happened" there and I responded to them at the moment. In Arizona I observed and recorded data without involving myself in the energies themselves. This proved to be a very unsatisfactory method, so I decided to personally submit to healing procedures in order to experience the effects of the force field first hand. For weeks I used acupressure, massage, meditation, and other balancing techniques. My initial reluctance to undergo treatments blocked any beneficial effects of the vibrations. Using ancient energy in archaeological research was one thing; experimenting with my body was another. My attitude began to change, however, after I had a healing experience that proved to be as satisfactory as any orthodox medical procedure. A small lump on the side of my neck had been dismissed by the doctors in Los Angeles as a sebaceous cyst. Its location made removal dangerous, but I was concerned about the size and potential danger. During a massage treatment I mentioned it to the healer. "Let me summon up the energies," she said, "and the lump will disappear." Having given up on other treatments, I decided to cooperate. She worked on the spot for twenty minutes, pressing gently, meditating, and moving her hand over the area. When I got up from the table, the lump was still there. Three days later it had diminished in size until it was gone. The next time I was in Phoenix, I asked her how she had performed the healing. "Simple," she said, "I changed the rate of vibration, and the lump dematerialized and vanished." "How simple," I thought, "I wonder if I could have done it myself."

A week later, after I left Tsaile, I came down with a partial paralysis of my face, known as Bell's Palsy. I thought at first I was having a stroke, but the doctors at UCLA assured me it was not a permanent neurological affliction. My own doctor estimated it would take at least six months to recover. At the very least, he said, there was hope for improvement in two months. "Two months," I said despairingly, "I have lectures to give in two weeks." I moped about the house, nursing my depression for several days. Then I decided to attempt a cure myself based on the methods I had observed in Phoenix. "What the heck," I figured, "I have nothing to lose." Each morning I stroked my face, setting the energies in motion, a difficult process in Los Angeles, where I always sensed a heavy level of vibration. I meditated, used incantations from Indian rituals, and pictured my face cured. I performed the healing ceremonies three times a day. Two weeks later the palsy had completely disappeared. I smiled a full smile, blinked my eye, which had been frozen at the outset of the illness, and observed the wrinkles that extended from the corner of my mouth halfway up my cheek. The recovery was a revelation and gave me confidence in my ability to make the energies work for me, rather than simply experiencing random encounters that I could not direct.

As part of the healing treatment for my face, I had spent a weekend in the desert at Palm Springs where Indian cultures once flourished. While relaxing in a whirlpool bath at the Spa Hotel, I saw a sign that said the bathhouse was built over a sacred well belonging to the Agua Caliente tribe, who used the waters for healing and religious ritual. Perhaps that was why I always felt renewed after leaving this resort.

I began to suspect that there were strong ancient energies in the springs as in the healing waters at Lourdes in France and the running brooks at Delphi in Greece. In fact, my research began to indicate that water was not only a conductor of energy but retained ancient vibrations at certain archaeological sites where its beneficial effects were used for healing, to facilitate control of the environment, and to enhance the power of sacred rituals. Dorothy Eady, an archaeologist, claims she could not see a thing without her glasses until washing her eyes in the well of Osiris at Abydos, Egypt; she never needed to use her glasses again.

Other religious groups had water cures as well. The traditional use of holy water, which apparently contained heavy charges of positive ions was one. According to Dr. Egerton Sykes, water was purposely used as a conductor of energy in the foundations of ancient structures such as the pyramids of Giza, and the Aztec temples at Tenochtitlan in Mexico City. Since water transferred energy and was a vital factor in the life cycle of many primitive people, they probably understood the rules that governed its behavior. Lowell Ponte, author of *The Cooling*, points out that native American tribes burned sweet grass and herbs, which, he believes, interacted with the ionization of the atmosphere to cause condensation and rain. During religious ceremonies conducted to influence the weather, American Indians may have consciously caused a change in the electromagnetic forces of the earth's surface. Days or weeks after the rites, odd weather effects were observed. I suspect that ancient people may have practiced similar energy transference and understood how to produce water by altering the wavelengths and molecular placement of energies. An example of the effect of human consciousness on weather is reported by Brad Steiger in *Medicine Power*, where he describes highly productive rain ceremonies carried out in Minnesota during July 1970.

In the Southwest, the energies of water used as a force for healing probably became more intense as a result of an increased physical charge of the minerals and the realignment of other particles as they interacted with human perception. Major religious groups may have copied the ancients by practicing baptism, which I believe also conducted positive energy through water to the newborn infant. Knowledge of these energy customs caused me to reexamine the tremendous power potential of the Southwest and consider how it might be tapped on a larger scale in planning future energy needs of the area.

A definite change of vibration could be noted in shifting environments from Arizona to California. In spite of the extensive Indian remains on the coast, the force field depressed my personal energy quotient. Except for the Chumash, who may have known the function of megalithic monuments taken from their former cultural connections in the Pacific, early California tribes did not actively reinforce the field. In California, the energy atmosphere appeared to be low and dense, possibly caused by the diverse energies of an influx of heterogeneous populations into the state. I suspected that in other sites, pyramids, monuments, and energy collectors maintained the vibratory flow in the absence of human channels, but California lacked all kinds of conductors. The state was like a toaster without a plug, and my experiences indicated that ancient energy demanded constant conduction to remain potent.

At a few cities in California I noticed heightened flow. On the highway leading to San Diego, I sensed a rapid shift in vibratory rate at the halfway mark, San Clemente. After experiencing the change on dozens of trips on this route, I searched for a principle to explain what I originally thought was a manifestation of my mind. In both San Diego and Los Angeles, there appeared to be no direct physical effect when energies grew lighter and more buoyant. Rather, it was something I perceived at an intuitive level which I believe was quantitative in form. Perhaps the energy itself did not change but only the interaction of the particles and the rate at which the wavelengths travelled. I believed every area had its own force field, which fluctuated according to the stimuli introduced into it, and the reinforcement directed at the mass itself. When I shifted environments from Los Angeles to San Diego, I discharged energy from one field and picked up the vibrations of another. At the midpoint, San Clemente, change began to occur. It usually took twenty-four hours to lose the old field and regain the new. Within a day, I was rejuvenated, refusing to think about Los Angeles so that I could stay disconnected from its energies. I believe the San Diego field was strongly influenced by the Mexican Indians whose energy practices originated with the Mayans. I had observed ancient Mexican rituals among the Yaquis of Guadalupe, Arizona, who fled persecutions in Mexico because of their bizarre religious practices. During Easter week, they held ceremonies of rebirth, reviving pagan practices related to energy and natural law. Tony and Sherry Arlotta attended the ceremonies one year, but Tony declined to take part in subsequent rites because he had absorbed the vibrations and become an unwilling channel for the powerful forces.

Tony, Sherry, and I had been pursued by members of this Yaqui tribe one Sunday when we entered their cemetery, also a Hohokam burial ground. Tony told me they were guarding the force field of the site and the spirits of their ancestors. Carlos Castaneda had Don Juan meditate with the Yaqui on his behalf since the tribe would not communicate with him until they perceived his state of consciousness. Other researchers reportedly faced death and expulsion by Indian tribes who refused to allow their sacred cities to be violated by disharmonious vibrations. For example, in 1924, Colonel Percy Fawcett disappeared while investigating areas of the Amazon, having violated Indian sanctions against the intervention of white men at their sacred sites. Contemporary tribes still guard the balance of the energies, forbidding dilution of the field by outsiders.

Modern man is totally separated from nature and the ancient traditions of prehistory. We no longer discern the possibility that cosmic energies affect all areas of life. Scholars such as Lyall Watson and Wilder Penfield have collected evidence for the existence of force fields, but skepticism and disbelief prevail. Few scientists acknowledge the reality of the phenomenon. Their casual encounters with the force have usually been insufficient to establish acceptance of the principle. Part of the problem is that consciousness determines an awareness of vibrations and this awareness is heightened through participation in the fields. Submitting myself to dozens of energy centers helped me comprehend the process. One cannot investigate the phenomena without becoming part of the essence itself, something most researchers are unwilling to do. Professors have told me they "believed" in the energies, but refused to become personally involved. I understood their reticence, since my early encounters with the force fields involved some pain, frustration, and a lack of consistent results. But after I understood the nature of the energies and took proper precautions in research, the work became a challenge. There were no rules or examples to follow. Ancient energy was a new ball game and I was not a very good player. I had to consciously make the energies work for me, as did Tony Arlotta, who was an expert in the techniques. Constant exposure to fields of force was essential, but faith was also an important factor. Even such a strong channel as Tony had to exert faith in order to maintain the flow when he immersed himself in the powerful Hohokam mounds.

I considered taking Tony to Egypt to do archaeological research there, but my plans for exploring the area were still in the formative stages. I also was concerned that his abilities might be reduced in an unfamiliar site. This happened to Dr. J. Norman Emerson, of the University of Toronto, when he took a psychic informant to Egypt. The channel, "George," successfully identified a series of Iroquois Indian carvings in Canada, stating the time, place, and maker of the artifacts without reference to traditional sources of information. Dr. Emerson reported the study at a meeting of the Canadian Archaeology Association in 1973, saying that George had received knowledge about artifacts and sites without the conscious use of reasoning. He presented the paper again in 1974 at the American Anthropological Association Conference in Mexico City, but scientists doubted that the research could radically change methodology in the field. Former astronaut Ed Mitchell explains such scientific backlash by observing that today's academic community, like that in the time of Galileo, refuses to look through the telescope for fear that the answers proposed to them might be right.

In Egypt, however, George was "blocked" and disconnected from the field, unable to receive information. Since many people are disharmonious in certain areas or fail to connect with them, I worried that Tony also might not function well in Egypt. Though he was balanced with Hohokam vibrations, his success abroad was dependent on his ability to relate to new energy fields.

My investigations of ancient energy in the Southwest encompassed a wide variety of experiences. The fields of force in and around Phoenix led to archaeological discovery and new techniques of healing. In areas of California around Palm Springs, I began to observe water as a powerful conductor of the vibrations. I learned, by contrasting California energies with those of Arizona, that the vibrations around Los Angeles were heavy and dense. These observations led me to new theories about human reception, conduction, and release of energy. Having observed and documented the phenomena in a wide variety of locations, my training and patience in studying the elusive forces all paid off. The Egyptian research proved to be more complicated, but my previous experience with ancient energy led to more definitive results since the experimentation produced concrete physical data.

4. Inside the Great Pyramid

My hunches about ancient energy were reinforced after investigating the Great Pyramid of Cheops in Egypt. The trip had been delayed for two years due to the unstable political climate in the United Arab Republic and the necessity of preparing well in advance for the rigors of experimentation at the monumental site. A small committee had been working with me since 1974 on the problem of ancient energy as it relates to prehistoric science and technology. Among the participants was Colonel Gerard La Rocca, an engineer with the NASA space program, and Bill Cox, an engineer interested in the phenomena of pyramid energy. The committee met on a regular basis and discussed their textbook research which included scale drawings and plans for experiments inside the pyramids at Giza near Cairo. These meetings resulted in a large volume of paperwork including reports on such diverse topics as microwave detection of ancient monuments, references to the construction of the megaliths, and detailed descriptions of archaeological sites from Alexandria to Aswan. Among our facts were bits and pieces of evidence suggesting that ancient energy could potentially be generated inside the Great Pyramid and that most pyramids possessed unknown numbers of hidden entrances and caverns where the ancients practiced secret ceremonies. Other discoveries indicated that the three pyramids of Giza may have been built thousands of years before the age of the pharaohs, and some hypothesized that the blocks may have been levitated into place by forces related to ancient energy. Unfortunately, none of the original committee members could spare the time or money to participate in the expedition. So, we agreed to recruit interested persons from all walks of life, the one requirement being that they share a common interest in conducting energy experiments at ancient sites. Twenty-five people were eventually selected from a large number of applicants. Ultimately, however, only a small nucleus actually took part in the experiments. An informal poll of the larger group indicated that "pyramid energy" was high on their list of priorities for the expedition. This was of serious concern to me, since I doubted the authenticity of such phenomena. In fact, I did not even connect pyramid power to ancient energy, since so many charlatans were promoting metal and cardboard pyramids and making outrageous claims about their capabilities. Makeshift pyramids, according to their promoters, could improve sexual prowess, resharpen razor blades, and keep yogurt fresh for indefinite periods of time. This commercial approach to a scientific subject tended to discourage me from investigating pyramid energy and its alleged modern uses.

One morning I picked up the newspaper and read an account of the work of Stanford Research Institute at the pyramid of Chefren in Egypt. Apparently, SRI researchers had attempted to penetrate the structure with an electronic device but claimed that excessive moisture in the atmosphere had negated any significant results. After trying for months to learn the *real* reason for the failure of their experiments, I discovered a friend who had an inside track to the project as his friend had been one of the principal financial supporters of the endeavor. He indicated that SRI, like the Russians before them, had been dismayed at their inability to successfully use traditional scientific methods, such as electrical devices, to secure data about the interior of the pyramids. In fact, the information from Egypt, when fed to the computers in this country, interfered with the recording device of the machinery and produced an inconclusive report. Russian frustrations and failures at the pyramid, described by Peter Kolosimo, in *Timeless Earth*, were similar to those of SRI as

The great pyramid of Cheops at Giza Egypt where the author explored three interior chambers to determine whether ancient energy techniques were used in the design and building of the structure.

reported in the newspapers. It was at this point that I became interested in the possibility that ancient energy could be related to pyramid energy. Perhaps ancient energy had been the undermining factor in the previous research.

With these considerations in mind, I arranged a meeting with Bill Kautz and Dolphin Lambert of SRI to take place in San Jose. The SRI officials asked me to invite Bill Cox, since they were interested in his experiments with pyramid energy. Bill and I spent four hours with them discussing their Egyptian research. They left us with reports and papers which convinced me that our upcoming investigation would be essential for shedding further light on their work and on my own theories concerning ancient energy. Although I continued a lengthy correspondence with Bill Kautz, his letters revealed less and less about the Egypt project. Finally, he wrote that SRI would cooperate with our organization on projects held in any country except Egypt, since the Egyptians were skeptical about the type of research we were conducting. Little by little our efforts at joining forces with SRI were disbanded.

Meanwhile, I concentrated my efforts on developing a means of measuring energies at the sites. The shifting nature of the phenomenon led me to question whether such measurements were even possible, but I was convinced that there had to be a means of testing the existence and manifestations of ancient energy. I examined all the conductors that I believed to be the triggering agents for the power. After eliminating color, light, and form as possibilities, I selected sound and vibration as the facilitating agents that would be the easiest to measure and observe inside the pyramid. Then I developed an experimental booklet and a kit to be given to each member of the search. The kits included such items as tuning forks, dowsing rods (specially designed with crystal tips), hand crystals, and harmonicas, and I also got a biofeedback temperature trainer to be shared by the group, a small machine, strapped to the wrist, which records temperature based on the body's reaction to energy as it enters certain subconscious areas of the brain.

A surprise conductor to everyone, including myself, was the flute. I included the flute after receiving a telephone call from Paul Horn just two months before our departure. Paul is a well-known flutist who reached international prominence through a recording he made inside the Taj Mahal in India called *Inside*. He had been trying since then to go to Egypt to record inside the Great Pyramid and was searching for a group like ours to assist him in his efforts. Although I had not intended to include the flute in our research, it became a distinct possibility after my conversations with Paul. He was not only a student of Egyptology and interested in all phases of our research, but he also had a real feeling for the question of ancient energy, possibly due to his training as a musician and the creative way in which he handled vibrations. With my assistance, Paul contacted many people and groups before our departure who were able to give him data about the acoustical properties of the pyramid chambers and the general construction and physical orientation of the monument. One of the members of our organization who had never been to

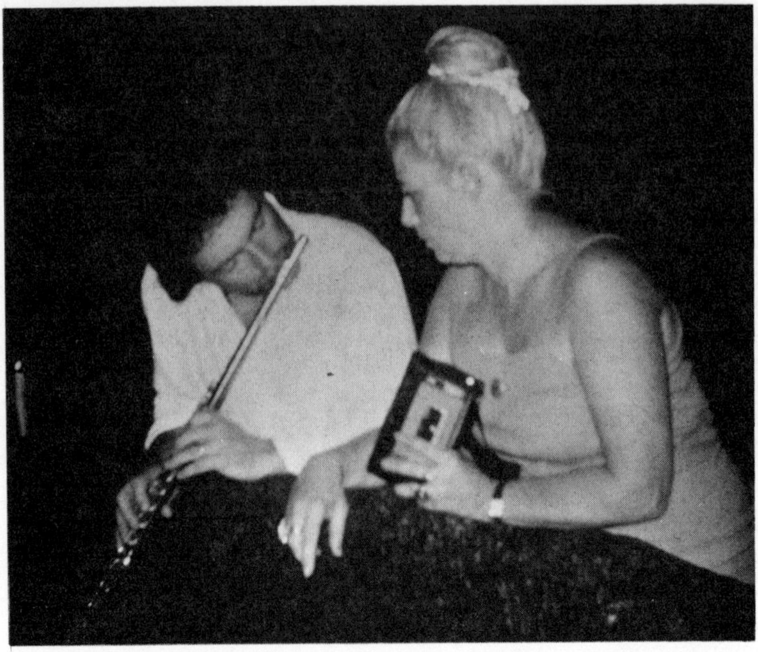

Musician Paul Horn tests the vibrations of his flute over the coffer in the King's Chamber at Cheops while the author records the amplified sound on a cassette player.

Egypt, Ben Pietsch, actually predicted the note Paul would hear when he tapped the King's Coffer with a tuning fork. In Egypt, Paul brought instruments with him into the pyramid to measure the sound exactly, and Ben's prediction proved correct acoustically and mathematically with a precision that amazed us all. This caused me to consider the possibility that Ben could tap into the information carried by ancient energy from a distance, since he was an intense student of Egypt and the pyramids.

Right up to the time of departure in May 1976, I continued to experiment with the most scientific way to conduct the research in Egypt and measure our results. I knew that I would be open to criticism from the academic community when I returned, a problem that had been plaguing me from the outset of my probes into the energy question. For years I had straddled the gap between two worlds, caught somewhere between the strict tenets of academia and the esoteric principles of the ancients. It was a difficult position to maintain and I found it hard to secure a definitive set of data relative to my theories. Yet I was cautiously optimistic that the Egyptian research might yield some physical proof. At first I inspected several machines that, their producers assured me, would be effective in measuring sound inside the pyramid. I later rejected these on the basis that the expense, weight, and general lack of reliability did not justify their use. Our preliminary experiments, it appeared, would have to be evaluated on the basis of observations and reports at the experiential level. Later, I reasoned, we could run statistical reports such as I had done in connection with my doctoral dissertation called Interrater Reliability, a process by which random numbers of people are interviewed, examined, or otherwise tested. The information or scores are then submitted to a statistician who plots responses given by the largest number of people on a scale. For example, of 25 participants asked to select the possible location of the missing records of Atlantis from a series of choices, 18 people might choose the pyramid complex at Giza while the balance of the answers could be spread evenly among the other choices offered. In evaluating the results of the test, the investigator would decide on the most promising place for the search based on the data received. Interrater Reliability is generally used for subjects too difficult to measure with traditional statistical tools. Since many of the participants in my expedition were anxious to conduct the investigations, it appeared there would be little difficulty in gathering a wide range of reports.

Such preconceived ideas about success inside the pyramid were shattered upon our arrival in Cairo. My earlier deep concern about working with groups became a reality. At least half the people expressed a desire to see the sights, shop, or sit in the sun rather than participate in the tests. Others declined to attend the daily training sessions, and still others wanted to "do their own thing." Paul, however, began his investigations immediately, and I went with him. The first day Paul, David

Greene (Paul's sound man), Roger Smeethe, and I went by taxi to the Giza area. To our amazement there were no tourists there at all, and we rented horses and rode around much of the complex. I was awestruck by the beauty of the vast desert and the visual impact of the three pyramids; they seemed almost out of place in the serene setting. After riding for over an hour, we explored the two smaller pyramids, Chefren and Mycerinos. We didn't have time that day for Cheops, however, since the rest of the group was waiting at the hotel for my return.

Early the next morning, we returned to Giza for our entry into the massive structure. Paul was not planning to record but only to explore and take photos. As we walked up the steep ramp leading toward the King's Chamber, I glanced up toward the ceiling and saw the enormity and precision of the blocks that comprised the passageway. They were the largest building materials I had ever seen. Normally I would have been dizzy and exhausted by the experience of climbing up the narrow, closed-in ramp in the hunched position required. Instead, I felt a sense of heightened energy and a light step, almost as if I were walking on the moon. The whole procedure of entering the King's Chamber had been effortless. Within minutes after we stepped into the room, Paul took out his flute and blew the first note over the King's Coffer. The four of us were alone in the chamber and the sound reverberated like a boomerang. I shuddered at the vibration and remained glued in my position on the opposite side of the room. My eyes moved immediately to the upper left-hand corner of the chamber behind the Coffer. Paul continued to blow notes at random. Thoughts about the ancient Egyptians came racing into my head. For some reason I had a mental picture of Paul standing in that same position thousands of years before. He appeared to be pushing a sheaf of rolled parchment into the corner of the chamber, using the power of the flute music to dematerialize the wall. I had heard of opera singers who could break a pane of glass several feet away with their voice. The two ideas seemed compatible but I chided myself for giving credence to what appeared to be a daydream. "Not scientific," I told myself, until I remembered the experience in the Etruscan tombs. After all, I had verified most of that data. Perhaps I would be able to do the same with the new flood of images from inside the pyramid.

Minutes later I walked over to snap a few photos of Paul playing. Four pictures later my camera jammed. That was the beginning of a long series of equipment failures and machines that refused to function inside the pyramid. When the final toll of broken equipment was in, our group had ruined three tape recorders, four still cameras, and one movie camera. The devices simply refused to function inside the chambers. Could ancient energy be powerful enough to have caused these failures? Perhaps what we thought were accidental happenings were really part of a larger plan which was working against our project. It may have been coincidence. Yet, I cannot believe it was. In his book *The Roots of Coincidence*, Arthur Koestler postulates a synchronicity in the universe and the possibility that there are no coincidences. He argues that events are drawn together by the harmonious action of corresponding vibrations and not through random occurrences. The Rhine experiments at Duke University also suggest that what we call "chance" is an unconscious product of energies which we still do not understand. Participants in the Rhine tests achieved success in causing certain number combinations to occur repeatedly in the casting of dice. Their results suggested to me that the events which took place inside the King's Chamber in Cheops were not accidental, but were part of a larger scheme whose rationale was presently hidden from me.

Paul, David, Roger, and I stayed inside the pyramid for more than an hour. I was filled with energy and moved from the King's to the Queen's Chamber with dexterity although normally it is a difficult route. It was amazing not to feel any of the residual effects of "Jet lag" from the trip or any lack of mobility normally experienced in the first few days in a foreign country. Instead, as a result of the pyramid explorations, I believe the energy of my mind and body was heightened, and I knew Paul, David, and Roger were feeling the same way. We were all in jubilant spirits.

For the next several days Paul made his arrangements to record a concert inside the pyramid while I returned to the group to select eight to ten people who seriously wanted to conduct the experiments. On the appointed day, ten of us entered the King's Chamber with the research equipment in hand. Since several of the participants were receiving independent study credit from the University of San Francisco for their efforts, they brought along notepads and other data retrieval forms to record the results. Each person worked with their own set of instruments. I chose the A tuning fork for my experiments. Wandering around the chamber, I struck the fork at various points along the wall and at the coffer. I could hear the vibration quite loudly but that was not an unusual effect. Then I struck the fork on the wall and leaned over to listen more carefully. Suddenly I felt a burning sensation on my ear. Grabbing a mirror from my knapsack, I looked at the ear. It was bright red on the lobe. I hadn't touched the tuning fork to my head so the burned area came as a complete surprise. Could the vibration from the fork struck inside the pyramid be powerful enough to physically damage my ear from a distance of two inches? I tried the other ear. Tapping the fork on the wall I held it away from my ear. Then I moved it closer and closer, still not touching the skin. The ear reddened and swelled. Later I tried the experiment outside the chamber. The effect was minimal. The acceleration of energy on the inside was incredible.

I remembered reading that the ancient Egyptians had not allowed people inside the main chambers of the pyramid on their first entry into the structure. The danger of going directly inside was mitigated by conducting rituals of initiation over a long period of time in order to increase spirituality and perceptual awareness. Afterwards, the initiate could penetrate to the core of the pyramid. Did the ancients know that physical and psychic damage could occur if people were exposed too quickly to the energies at their point of maximum potency? I considered this information important and related to the injury on my ear. Perhaps even I was not prepared for the intensity of the experience inside the King's Chamber. But I was mindful that not everyone is affected in the same way. Later I saw groups of Germans and Italians who entered the chamber, looked around, took some photos, and left. It seemed to take a certain kind of *receptive* individual to be affected by the ancient energies.

Still pondering the tuning fork experience, I turned my attention to the other members of the group who were performing experiments with their own instruments. One of the women was holding the dowsing rods straight out in the corner behind the coffer. Although her hands and arms were still, the rods were bouncing up and down as if a stiff wind was blowing. She was amazed. This was her first experience with the energies. Before we could proceed further, the lights went out in the chamber and we all stood in the darkness for at least five minutes. Fortunately, the inside of the pyramid maintains a cool 68-degree temperature so we were not uncomfortable. When the power was finally restored, we conducted the rest of the experiments. We recorded the harmonica sounds and the vibration of the other instruments. We also dowsed in other locations in the chambers. Everyone was aware of the energizing effect of the chamber, the heightened vibrations of the instruments, and a general increase of awareness and perceptiveness after spending close to an hour inside the pyramid. Yet, it was too soon to evaluate just how seriously we were affected and whether ancient energy was responsible for the results. I knew that we would need to replicate the experiments dozens of times under controlled conditions to ensure the accuracy of our reports. But would standardized scientific procedures work in connection with ancient energy? Such instruments and operations were adequate for measuring phenomena with a strict cause and effect relationship. Verifying test results concerning ancient energy, with its other-dimensional components, might require a different kind of standardization. Could scientists be trained to develop such tools? I felt confident that with time and training the scientists of our modern age would measure up to such a task. Yet at this point, none of our experiences could be accepted as fact.

The small group was still highly motivated after the experiments, so we decided to chance crawling underneath the pyramid where a smaller chamber exists lying on a diagonal beneath the earth's surface. This chamber is less than half the size of the King's Chamber and has a structure inside called the "well" which appears to have no bottom. The few visitors who pay off the guards in order to enter the room have tried to drop stones or lower ropes into the well's opening in order to find out its depth. To date, no one has been successful in finding out where the opening leads and what the function of the well was in the ancient world. The underground passageways are generally closed to the public, but we gave the guard a sufficient tip or *bakshish* and he opened the door. One couple decided not to venture underneath, so eight of us were left. Using flashlights and candles, we crawled on our hands and knees for at least half a city block. There was barely six inches of space above our backs as we moved along in the dust. Normally, my claustrophobia would have prevented me from continuing the trek, but in this instance I hadn't the slightest hint of dizziness or lack of air. We just all kept going, propelled forward by our enthusiasm in that strange environment. About three-quarters of the way to the chamber, one of the men screamed, "Help me I think I'm having a heart attack—I have to go back" A second man voiced similar concerns, so we all flattened ourselves out to let them pass by, and they crawled in the direction of the entry. Neither man had ever been seriously ill before, and I learned later that they were completely recovered within minutes of leaving the pyramid. The experience was strangely reminiscent of my own Chichen illness in the Yucatan. Had the two men been been affected by the energies that some researchers believe are heavily concentrated at the base of the pyramid? An archaeologist working in the Yucatan had explained to me that the Mexican and Egyptian architects may have known how to place a reverse pyramid shape under the ground beneath an upright pyramid to form a geometric shape known as a tetrahedron. This tetrahedron vortex would cause the energies to be concentrated at the center of the two pyramids which, in most cases, was at ground level. Paul and I had seen a Danish film of the inside of the pyramid before we left Los Angeles that showed tunnels beneath the ground at Cheops not identified in any of the standard texts. I wondered if these passages led into a reverse pyramid, but there was no way to verify such an idea without demolishing the pyramid itself. We also had been told that the energies were more heavily concentrated at the top of Cheops, but our later experiments did not bear this out.

It took us about fifteen minutes to reach the inner chamber, which was totally different from the King's and Queen's chambers in appearance and size. The room was small and had no megalithic blocks. There was a well-like structure at the entrance which dropped down into an endless pit. I began to examine this opening when suddenly I had the same feeling the two men had experienced—I had to get out. Some of the group wanted to stay and meditate, so I left them there and moved on my hands and knees with all haste through the passageway toward the door. When I emerged, I discovered that my new tape recorder had broken as I was recording conversations in the inner chamber. It never worked again. The mechanical failures were devastating. The day before, the movie camera had broken, which cost Paul and me a tidy sum since we had brought along a sun gun, other lighting equipment, and film to shoot a 16mm motion picture of the experiments.

Back at the hotel that night, our group evaluated the day's work. They discussed their experiences in conducting the tests and agreed that the pyramid had affected their awareness and perception. Naturally, I was anxious to collect data retrieval forms from other expedition members who were conducting research independently from the group effort. Since many of them were still performing experiments over a period of several days, I did not press them for details until we arrived in Luxor some days later. (For detailed description of the experiments see Manual, Appendix D.)

Exhausted from my investigations, I booked an appointment with the hotel's masseur, Abdel Fetah. This was a perfectly proper procedure, since the ladies were advised to wear their bathing suits during the treatment. Abdel spoke English fluently; we were able to understand each other with no difficulty. Before long I was discussing the problem of ancient energy with him. After about half an hour, he asked if he could share a very private experience with me. He said he had been going to the Giza area ever since he was a young boy. For some reason he had been drawn to the site and claimed that he could actually "see" through the ground when he was there. He also reported that he observed tunnels underneath the sand connecting the Sphinx to the Cheops pyramid as well as doors and chambers near the Valley Temple to the west of the Sphinx. I shuddered when he said that, since the great psychic Edgar Cayce had predicted that the records of Atlantis would be found in this location, buried underground in chambers near the paw of the Sphinx. Abdel had neither read nor heard of the Cayce books and the similarity of the accounts was incredible. He also claimed that the Giza site energized him to such a high degree that he felt "out of control" when he went there. In fact, his channeling of energy disturbed him, and he often tried to "break the circuit" of this force before he came in contact with other people. He accomplished this by eating a heavy meal or by taking a bath or shower. I suggested that perhaps he could spend some time with our group, and he made plans to sleep on a flat surface at the top of Cheops pyramid with four or five of our members. Although we all assumed that some miraculous energy effect would take place at the summit, he told me later that the experience had not produced any effect on him whatsoever. In fact, he said he was bothered by having other people around and he could not make the energies work for him that night. I fully understood his predicament, having frequently experienced a blockage of my own energies in large group projects.

Four days after our arrival in Cairo, Paul finished recording his all night concert in the King's Chamber at Cheops. Since I was anxious to learn how things had turned out, I approached him cautiously with my questions. Musicians are often temperamental, and Paul was no exception. Since he had demonstrated little interest in the psychic component of the energy question, I was careful not to phrase my questions to encompass any sensitive areas. His response to my inquiries, therefore, came as a surprise. When I asked him what happened, he said, "You'll never believe this. David and I actually thought we heard voices in that chamber. We also heard sounds that we could not identify and notes that I knew I had not played." Then he briefly described the intense experience inside the room and invited me to hear an unedited version of the music that was recorded that night. I closed my eyes and rested while Paul turned on the recorder. The sound blared out as if it was Judgment Day, pouring an intensely rich and forceful vibration into the room. The melody overtook me and I enjoyed the concert immensely after so many hours spent in research at the Giza plateau. I felt Paul had truly captured the essence of the ancient structure, yet the music was strange and unfamiliar, with varying levels of monotonal sounds designed to induce a meditational state. It was only later, when I heard the entire album played in the United States, that I fully appreciated the magnificent composition Paul had created in his album *Inside the Great Pyramid*. Paul and I barely spoke again for the next six days. For some reason (at a very deep level of awareness, I believe), the experience inside the pyramid caused him to back off from me. After serious contemplation and discussions with David, I felt that Paul may have wanted to release the effect of the energies inside Cheops. Since I was a strong channel for that force, his instincts kept him away from anyone who might expose him again to that powerful vibration. After a 10 day stalemate, our relationship returned to normal again on the flight back to Los Angeles.

I had observed other people open up to the energies and then close off when the forces became too powerful. Abdel was faced with that problem, and I had heard that the famed Uri Geller monitored his own strength in the same way. For example, one member of our group never discussed the trip with his colleagues after he returned and dismissed anyone who asked him about it. We had lunch together one time after the trip and he said, "That pyramid experience—I'd rather not discuss it." Another woman, a professor of anthropology, refused to enter the pyramid at the Giza area, claiming her ill health prevented it. At the outset of the expedition, she tried to conduct an experiment on her own but she was perplexed over the energies, demanding that I demonstrate the physical existence of the phenomena, which I could not do for someone else. Her own inability to handle the vibrations frustrated her investigations. The harder she tried, the less she produced. Eventually she learned to experience an energy effect by working with another team member dowsing the pyramid area near Sakkara but she remained a skeptic. Roger Smeethe and Susan Hazlewood, two powerful channels in the group had experienced the opposite effect. They decided to climb the Cheops pyramid, and advancing one third of the way up, a powerful force pushed Roger down to the ground where he struck his head on a large rock. Susan later told us that Roger did not lose his footing; he was an experienced climber and athlete. Something had happened on the way up to the top causing a loss of energy and a shift in awareness, probably due to Roger's changing consciousness. I was reminded that the Navaho Indians placed their healing centers one third of the way up their pyramidal structures. Since the King's Chamber is also one third of the way up the pyramid, Roger may have been affected by the stronger energies even though he was outside the structure.

After I returned from this expedition, I collected data and searched for clues related to the ancient energies we had all experienced. Since no common thread or basic explanation tied the information together, I decided to reexamine the people who had participated in the study. This change of direction produced some interesting facts. It appeared that most often a person with advanced academic degrees was less successful in understanding the energies than the layman participant. I also noted that members who I believed were highly developed at a spiritual and/or psychic level experienced a greater degree of success in their experimentation. It was also interesting to observe that members involved in the creative arts, music, or communications were more favorably disposed to the experiments than the professionals in our group such as teachers, nurses, and lawyers. Finally, members who were emotionally inclined were sometimes incapable of handling or understanding the energies. I had noted this problem on previous expeditions, and therefore I was careful to monitor the feeling level of the people who submitted to the energy experiments. The power of the sites could accelerate mental and emotional problems that I was not prepared to handle as the expedition leader. The positive uses of the energies inside the pyramid were easier to manage. Energetic, happy people felt "high" after being inside and remained stimulated for hours afterward.

The observations made in Egypt were not conclusions about the nature of ancient energy but rather a collection of experiences which I used as baseline data. They did not "prove" the existence of ancient energy but were a catalyst that encouraged me to continue exploring other sites and structures to verify and explain what had happened. The additional data to support my theories about ancient energy came in a series of investigations at Abydos, Luxor, and Dendera along the Nile conducted in a second research expedition in October of the same year.

5. Powerhouses of the Nile

The tombs and temples of Upper Egypt were closely linked to Egyptian religious life. Color, sound, and light were used by the priests in ceremonies where the initiate learned to contact higher thought forms. The rites were similar to modern revival meetings where the faithful receive God in an atmosphere designed to induce religious fervor. My curiosity over the possible connection between Egyptian religious practices and ancient energy brought me to Luxor, where our group included the well-known dancer and actress, Ann Miller. Ann and I engaged in an exploration of the Valley of the Kings, since she had an intense interest in the pharaonic period of Egyptian history. Ann had started down the Nile on three previous occasions, but some obstacle always aborted the trips. She seriously believed her mishaps were caused by an ancient curse. She also believed she may have been the reincarnation of Queen Hatshepsut, whose temple faced the Nile at Luxor. Initially, I was amused by Ann's statements, although I had an open mind about the possibilities of life after death. The work of Dr. Ian Stevenson at the University of Virginia and Dr. Elizabeth Kübler-Ross in Chicago suggests that reincarnation is an idea to be seriously considered. Curses were also well known in connection with ancient sites, partly the result of stories about misfortunes suffered by Howard Carter and his team at the opening of King Tut's tomb. Perhaps Ann was able to contact

Actress Ann Miller on an expedition into the Valley of the Kings to find clues to the mysterious Egyptian Queen Hatshepsut. Miss Miller believes she may be an incarnation of this famous Egyptian ruler.

energies from the past. The idea was fascinating but impossible to verify.

The assistant mayor of Luxor, Aboul Magd Mahmoud, arranged our visit to the Valley of the Kings, officially appointed by the municipal archaeology department. A personal guide accompanied us across the Nile, delivering a running commentary on the history of the site and its importance during pharonic times. I asked him why there were no pyramids in Upper Egypt, since ancient Egyptian religious leaders used the interior chambers of pyramids as initiation centers in the passage to life after death. He answered that the tombs and temples in the Valley of the Kings may have served the same function as the interior of the pyramids. My experiments inside Cheops had led me to believe that knowledge of ancient energy was known to the builders at Giza. Could the tombs and temples of Luxor also be in harmony with the timeless force? If so, the pharaohs at Luxor had had access to the same advanced concepts used in the construction of the great Pyramid. Thus, the civilization that once flourished in the Valley of the Kings may have left a force field of considerable intensity, still manifested in the existing structures. Ancient priests would have learned to activate the energies and use them in the ceremonies of initiation where the participant dealt with the transition from life to death. I was still pondering such ideas as we travelled along the dusty road toward Nefertiri's tomb.

Formerly the tomb was open only to archaeologists. We were the first lay investigators allowed inside after the initial excavations. The magnificent structure contained panels of paintings in a brilliant range of colors. There were elaborate symbols on the walls as well as openings for the flow of light. After a few minutes in the first chamber, I became aware the colors were affecting my perceptual field. I experienced some dizziness and a slight altering of my timespace perspective. I recalled reading a report by Lyall Watson in *Supernature*, which stated colors are powerful stimulants in the raising of consciousness, since they have textures which could be physically experienced. He had mentioned the Luscher Color Test of twenty-five hues which produced a variety of feelings among its subjects. Participants picked up the frequencies of the colors and translated them into widely different reactions. I also remembered that Paul Brunton, in *A Search in Ancient Egypt*, inferred that the ancient Egyptians understood the principles of structures, symbols, and colors and used them in the ceremonies of initiation. They covered the eyes of some subjects during religious ceremonies, perhaps to allow for the reception of energy at a higher level of awareness and response. That day I believe I became an initiate, without even trying, when the lights went out in the second chamber. As I clung to the frescoed walls trying to regain a measure of sight, I still "saw" the colors from the first chamber. I even felt their effects. The stimuli, after reception, apparently entered other areas of my central nervous system. I was confident that I had perceived the colors but the perception came from a sense other than my eyes.

Nefertiri's Tomb in the Valley of the Queens, Luxor Egypt. Color, form, and design inside this tomb may still create an energy effect, distorting the visitor's perception.

The colors and designs in the second chamber were muted by a gauzy substance draped over the murals to prevent moisture and dust from destroying the hues. I moved my hand over the covering. It glowed with an eerie intensity that brought to mind the ancient ceremonies that had taken place here. The colors flowed one into the other as in the broad spectrum of a rainbow, giving the feeling of illusion and movement. I shuddered at the sensitivity of the colored frescoe that covered the wall. I was as vulnerable to color and form as the ancients. Itzhak Bentov, in *Stalking the Wild Pendulum*, points out that rhythm in color and design can affect us biologically and exert a gravity-like pull. My altered perception in the tomb may have been due to the magnetic effect produced by the colors. I was drawn to them in the same way that I had been "pulled" towards objects and inscriptions in the caves of the Mediterranean and the Etruscan tombs.

Most monuments in the ancient world were brightly painted, including some of the now white Greek marble statues and outer coverings of Egypt's tombs and temples. Sacred initiation and religious rites partially depended on bright colors for stimulation, and some of these hues have never been duplicated. An example of an unduplicated formula for color is "Pompeiian Red," found on the walls of the city of Pompeii. The ancients may have created and preserved color based on the knowledge of a balanced vibrational field. In *Quest for Meaning*, the ABC television special, 1978, it was suggested that artists prior to our modern era painted in accordance with Newtonian physics strictly reproducing observable physical reality. Later, cubist artists such as Braque and Picasso liberated the space-time concept. Their work was based on the idea that the experience of the painting was a function of the mind of the viewer. The ancients, like the cubists, used the principles of energy governed by the laws of relativity. In *The Neutrino*, Isaac Asimov affirms that early cultures perceived the laws of nature as different from the laws of the universe. One set of principles operated on earth and another functioned in the heavens. Ancient colors may have been created according to a set of principles which did not distingush between material and nonmaterial reality. By recreating such ancient techniques, we may discover a new way to preserve our own artistic and cultural heritage. It is possible that historians could learn to record thoughts and ideas by developing symbolic forms of energy to be decoded by later generations. In other words, the heritage of our civilization could be stored in a set of symbols universally understood by men in all ages. This would require a new cryptic system by which records could be preserved. For example, babies communicate in non-verbal concepts. The infant utters a small sound, accompanied by a certain body movement, and the mother "picks up" the message as accurately as if the baby had uttered an entire paragraph. I believe such limited verbal communication can potentially carry more impact than stacks of printed pages, if the verbal symbol is charged with energy carrying a conceptual message and the communicating parties are receptive to the interpretation of the symbols.

The kind of historical record-keeping I am suggesting implies a new awareness by which people in the future would have to be open to these techniques and able to evaluate the kinds of symbols used in such a system. The code would also apply to documents such as the *Dead Sea Scrolls* and the *Egyptian Book of the Dead* which would take on new meanings, both literal and symbolic, since these early writings were developed with a hidden as well as an obvious message. A special code was also incorporated into the Bible along with traditional meanings. The Cabala, mentioned earlier, was a code developed by a Hebrew religious sect known as the Hassidim, created in order to understand and interpret scripture through numbers and symbols, but later it became a mystical system used by other religious sects and secret societies. Its main tenets are embodied in the *Zohar* or *Book of Brightness* which sets forth a universal plan of symbols known only to the initiate. An example of a cabalistic approach to a subject other than biblical study is a new field of investigation called mathematical musicology where symbolic number systems give another dimension to the basic melody. The ancient Egyptians may have enhanced the vibration of temples and tombs by conserving knowledge through colorful symbols and shapes painted on the walls. I strongly feel we cannot fully understand the history of the ancient world unless we learn to decode the messages hidden in code which probably express the deepest feelings of the people.

As I left Nefertiri's tomb, it seemed to me that the physical imbalance I had temporarily experienced was due not only to the altered state of awareness produced by the colors and symbols, but was a natural result of the swift change in the density of the vibrations. A shock factor can operate at ancient sites due to the rapid switch from one force field to another. Sensitive people such as myself are more vulnerable to such experiences, which are alleviated by continual and gradual exposure to the sites.

After Luxor, I moved on to explore the energies at another sacred site, Abydos. I had been to the area earlier in the year, searching for clues to the problem of the Atlantis records—documents believed to contain a history of that lost civilization—which Edgar Cayce states were buried in Egypt. Since the trip was long and arduous, I did not look forward to revisiting the ruins. I was exhausted from the problems that had befallen our group, although the time spent with Ann Miller helped to distract me from the difficulties I had encountered as the leader of the expedition. The Lebanese War snarled travel arrangements, participants were incompatible in personality and temperament, and I was ill during the first four days of the journey. Up to the time we arrived in Luxor, these distractions had prevented me from focussing on the problem of ancient energy and depleted my normally positive attitude and high physical stamina, important to the success of such investigations. After Luxor and the experiences at Nefertiri's tomb, however, my attitude changed. I had a renewed interest in exploring Abydos and learned that it may have been the oldest city in Egypt. The site is located on the Nile not far from Luxor, and is remembered as the home of the legendary Isis and Osiris. The river itself had been viewed by the ancients as a deity since it nurtured the soil and revitalized the ancient Egyptian world. In the religious ritual that grew up around Abydos, the Isis and Osiris legend was told and retold, often enacted in passion plays. According to oral history, the body of Osiris was hacked to pieces by his jealous brother Seth and scattered to the four corners of the land. Isis, the wifesister of Osiris, recovered the pieces at widely diverse sites, including Abydos, Sais, Dendera, and the Faiyum Delta. Later Isis was impregnated by the repaired corpse and produced a son, Horus. The myth is important for the study of ancient energy since the disbursement of Osiris's body throughout the land may have represented a fertilization or replenishment of energy in the soil. I believe it may also have a basis in fact, since mythology states that Isis and Osiris waged a prehistoric confrontation in the struggle for the unification of Upper and Lower Egypt.

In my opinion the energy created by burying the parts of Osiris' body can be compared to the power of the ley lines scattered throughout Britain, Ireland, and Wales, reported by researchers such as John Michell and Francis Hitching. Ley lines are ancient tracks in the earth, often found near mounds, unworked stones, moats, and crossroads. Recently, Michell and others discovered that these lines are also sun alignments, built by the ancients, which serve as conductors for the natural energies. The Romans and other conquering tribes, aware of the power at certain sites, frequently built their road systems on top of these markings. It has been suggested that these points also correspond to the acupuncture centers of the human body. Since the head of Osiris was found at Abydos this may signify that the city was the most potent and concentrated arena for energies, and of paramount importance in religious ceremony.

The ancients could have used mythology to raise consciousness and teach the principles of natural harmony and balance. Their function was a kind of "attention getter" like myths, colors, mandalas, and crystals, which awakened higher centers of awareness. The anthropologist Claude Levi-Strauss emphasizes the importance of myth structure in the ancient mind which was apparent at Abydos where the myth of Isis and Osiris was painted, portrayed, and narrated throughout the temples to remind the people of their human responsibilities and their place in the universe. During a religious ceremony and at the initiation rites, the wall panels must have been stimulants for the participants.

Abydos was constructed in a series of layers, and excavations exposed the remains of countless ancient cultures. A cutaway of the hillside revealed megalithic palaces with post-and-lintel construction, dolmen architecture, blocks set in place without mortar, and underground tunnels stretching for miles alongside the Nile. Due to lack of funding for archaeology in the Egyptian government, the interior of the tunnels had not been scientifically examined, and little was known about the history of the site. After a long climb down, I arrived at the level of the river and peered into one of the squared-off tunnels. Just as I began to crawl in, a resistant force or hunch caused me to back off. I had obeyed such premonitions before when my intuition indicated that it was unsafe to continue explorations. I always followed these feelings just to play it safe. Perhaps my own consciousness was out of alignment, or the vibration inside the tunnels was heavily negative, like those I had encountered in Yucatan. I remembered reading in *The Dream of Troy* by Arnold Brackman that the archaeologist Schleimann suffered physical and emotional discomfort at each of the nine layers that led to the discovery of Troy. On several occasions serious illness, poisonous snakes, dangerous insects, and other obstacles had halted the excavations. Archaeologists such as Schleimann frequently deserted digs for reasons other than lack of money or time. Such abandonment may have resulted from powerful exchanges of energies at the sites, forcing workers to discontinue projects such as the excavations at Pompeii or the recovery of temples and pyramids near Tikal in Guatemala. Archaeologists who were not harmonious with the energies of the area often suffered work disruption and other difficulties.

The myth of Isis and Osiris continued to plague me, since I believed some underlying principle accounted for its longevity among the ancient Egyptians. The story was not only connected to religious ritual, but also demonstrates a balance of energy that resulted from sexual exchange. Isis and Osiris personified the male-female equation, or twin spirit relationship, mentioned in ancient documents. Sexuality is an underlying theme in the Song of Solomon, although theologians have given other interpretations to the documents. Gnostic codices discovered at the caves near Nag Hammadi reveal sexually explicit passages connected with the rites of early Christian mystics. Sex used with religious ceremony was encouraged by some cultures but generally condemned by others. The Bible has sanctions against such practices, yet sex and religion were connected in sacred ceremonies at Malta, Crete, and among other civilized tribes of the ancient Mediterranean world. The loving relationship of Isis and Osiris produced an energy which served as a creative stimulus throughout Egypt. Their polarity symbolized the flow of energy which harmoniously united the country. The vibration created by perpetuating the duality principle pervaded the sites at Abydos and Dendera, where the overriding feeling was unity and balance.

Dendera was a short ride from Abydos, but the city differed in construction and vibration from all other ancient cities in Egypt. Legends relate that Isis and Osiris took flight to Dendera from an ocean-based land in 16,000 B.C., where they constructed monuments and temples in commemoration of their homeland. Giant shells fanned out on the walls of the temples and were also imbedded on the tops of the pillars. Astrological ceremonies were the main focus of attention in the temples. An opening in the dome allowed the sun's rays to penetrate the interior. I believe the rays may have been used as consciousness-raising devices, since light, like color, has a vibratory effect on human perception. Marilyn Ferguson theorizes in *The Brain Revolution* that the Egyptians may have used the sun as their point of focus for entering into the meditational state. The peephole was the stimulus that triggered consciousness away from the material world and toward the universal force.

The entire zodiac also preoccupied the ancients at Dendera. The famous plate of Dendera, found at the complex, contained primitive symbols of the twelve astrological signs, represented by human and animal configurations. These signs were also used to evaluate human behavior, since the ancients knew that shifts in consciousness were related to the influence of planetary energies on the human psyche. Originally, the signs were used as part of the ritual of initiation in religious ceremony. People entered the temples and were drugged or hypnotized and thus rendered inert, almost lifeless. The soul was coaxed out of the body and urged to contact the "higher" forces. Through this method the initiate learned to adjust to the moment of death when spirit (energy/consciousness) left the body and flowed on to join a higher source called the *Nedyt*, or unknown. In modern terms, the book *Life After Life* by Dr. Raymond Moody describes the experiences of people who physically died and were later revived. Their stories are similar to reports of the ancient Egyptian initiation experiences. In both cases, the subjects remembered certain sounds, images, and vibrations as they moved in consciousness from the physical to the nonphysical state.

In the ancient ceremonies, the priests assisted the subjects through a sort of rehearsal of the experience of death so that the feeling would be familiar when death actually occurred. Thus ancient Egyptians were assured of immortality since they could physically witness the transition through their own sensory processes. The condition of the initiate can be explained with reference to Einsteinian theory which states that the reality of density and weight are determined by man's ability to perceive matter as physical form. An object, therefore, does not actually have a solid state, nor does it occupy a particular space, but is rather a series of events reconstructed in the human mind. Ancient Egyptians, long before Einstein, understood these principles and practiced them in religious ritual and ceremonies of initiation.

I was less successful in encountering ancient energy at Dendera than I had been in Abydos, since the overexuberance of my companions deadened my own sensory field and powers of concentration. However, I did perceive an order and balance which may have resulted from the religious rites that had taken place in the temples thousands of years before. In quantum physics we observe that order can come from a disordered group of subatomic particles. During the initiation rites, powerful energies may have been emitted from the consciousness of the subjects. These energies were naturally disharmonious, having varying wavelengths. I believe such vibrations remained in and near sacred sites like Dendera where they eventually were transformed into a more harmonious state. Ancient energy always seeks to regain balance, and eventually does, if it is not interfered with. Thus, through the ages, I believe the disordered vibrations of the temples and tombs emanating from religious rites were transformed into the stable energy presently perceived at the sites.

Abu Simbel Egypt. The author's research indicates that pyramids may once have existed at this site before the statues were moved into their present location.

One of the highest energy centers in Egypt is Abu Simbel, an archaeological site south of Aswan where statues were relocated when the dam was built. Here I observed the tremendous feat of engineering by which the giant structures created by the Pharoahs were transferred to higher ground during the building of the dam. Egyptian archaeologists on our team told me that they believed other cultures once flourished at Abu Simbel though no evidence of such civilizations had been unearthed. There were five documented Nile Rivers, the last of which came into existence three million years ago, a date which coincides with the Leakey finds of ancient man at Olduvai Gorge in Africa. Perhaps Abu Simbel in its strategic location along the Nile had been a site of high civilization in many remote areas of the past.

Shortly after I entered the temple, which was inside the confines of the giant statues, I heard someone shouting my name. To my amazement, one of our team members, an anthropologist from South America, was holding a polaroid picture in her hand, still wet from the developing solution. Maria was ecstatic. She had tried her luck with the energies at the pyramids of Giza but had had no measurable success. She was disgruntled at her failure, and she blamed me for not giving her adequate instruction, though I assured her that the power was a product of her own consciousness. Now, at Abu Simbel, the energy finally manifested itself to her in physical form. Carefully focusing the camera, she took what she hoped would be a stunning shot of the giant statues. She developed the picture and what did she get but an image of the pyramids. I tried to find out if, by accident, she had inserted a partially used roll of film from Giza but she assured me that she had put in a new roll just before we left the bus. Everyone in the group observed the picture, yet none of us had an explanation. One possibility occurred to me. Television producer Alan Newman had witnessed a Japanese boy projecting a thought form of the Tokyo Tower into a polaroid camera. The tower appeared on the film moments later. Could Maria have concentrated on pyramids while photographing the statues themselves, dematerializing one and materializing another? Another possibility was that pyramids may have stood in place of the statues in times past. If so, they might still have been in a supersensory range of human perception, beyond the visual field. When Maria took the picture, the pyramids may have rematerialized on film by a process not fully explainable at this time. I believe this may explain why some people see UFOs, report disappearances into the Bermuda Triangle, and obtain unusual effects inside the pyramids. In these situations people probably interact, consciously and unconsciously, with the ancient energies. They are able, through their own perception, to see beyond the level of causal reality to spheres where time has no meaning. Ultimately, they alter their own reality, making objects appear beyond the causal plane. Experiments in Kirlian photography have shown that after cars leave a parking lot, their energy fields remain intact. Shadows of the cars have been seen on film. Thus, though the physical form of the pyramid was gone, the image could be perceived in the plane of higher sensation.

Maria became less of a skeptic after the film incident. She acknowledged the evidence that ancient energy had a basis in fact even though it often behaved in strange ways. It was the erratic nature of the phenomenon, and the lack of replication, that caused denial and disbelief among academics. It took a powerful physical demonstration to convince Maria that ancient energy was worth investigating. In a sense, the early Egyptians gave similar demonstrations to their initiates. They taught them to trust the "other world" by giving them a tangible experience with immortality. After the ceremonies, the initiates were able to perceive two systems of energy existing concurrently in the universe.

Many tribal people have an innate awareness of nature and energy. The Nubians, who live in Aswan, are extremely sensitive people. Their musical instruments, especially the small stringed lyre made of wood, metal, and animal skins, are plucked with one finger. Once, after listening to this sound I asked the player about its unusual piercing vibration and high frequency. He explained that Nubian music bridged a gap between the reality of this world and the natural harmony of the cosmos. The listener, or the player, tuned into higher consciousness through music. Jewelry was also a part of the Nubian ritual. It was designed to give the illusion of motion, and was used in religious ceremony, for decoration, and in dance. Ancient Egyptians had broad jeweled collars, as did the Nubians, which I believe were worn to encourage the energy flow. Today tribes such as the Maori of New Zealand still inscribe signs and symbols across their faces and breasts to harmonize more closely with the universal force. Photographs of these body tattoos appear in *The Mystic Spiral* by Jill Purce.

I left Egypt with a high regard for the science, wisdom, and technology of the people who had lived along the Nile. At every site, ancient energy was reflected in religious ritual, art motif, color, and mythology. Somehow I felt linked to the cosmic rhythm experienced by the ancient inhabitants of Egypt. The concentrated energy of the tombs, temples, and sacred sites allowed me to fully sense the higher vibrations and to come to a clearer understanding about the characteristics of the people who had lived along the Nile thousands of years ago. I was also better prepared for chance reactions to the sites after my earlier encounters with force fields in the Mediterranean and the Southwest. In short, my expertise at managing the energies had improved as a result of the second Egypt expedition, which also provided twenty people with the opportunity to sense a new phenomenon first hand. For the beginning investigator of ancient energy, Egypt is one of the most powerful settings anywhere in the world—an archaeological wonderland where I believe the energies have only started to reveal their true potential for the advancement of mankind.

6. Releasing the Force

Energy is neither created nor destroyed. My father explained this concept to me in response to my concerns about death and the hereafter. He assured me that particles of matter change their life span and length of wave but that the substance is never lost. At that time I was too young to perceive the qualities of an invisible essence. I never forgot what he told me, but my perception was still firmly grounded in physical reality. What I learned through my five senses, I accepted as fact. Anything else was speculation. My father said that some energies, apart from physical forms, were timeless and endless. This idea also boggled my mind, since I could only imagine finite laws that followed the principle of cause and effect. Later, as I began to investigate ancient energy, I recalled the characteristics of relativity he had taught me as I continued to question the nature of the matter. Did it move around? Did it circulate through people and things? Energy in my home was transmitted and received via cords, sockets, and outlets in the wall. Power came from huge generators in local plants. The stove and refrigerator ran on electrical current unless the socket overloaded. But the energy my father described was not as easy to explain. For one thing, I couldn't locate the transmission mechanism or source of the new flow. Electricity was observable. When the fuse blew, parts of the equipment were destroyed.

Ancient energy was elusive. I sensed its power over people at archaeological sites, but there was no consistent pattern to their experiences. They spoke vaguely about feeling "high" on Crete or depressed in the Yucatan. I was convinced from my investigations that most ancient cultures constructed their monuments over powerful fields of energies in the earth. I suspected that the potency of these areas caused an accumulation of energy in their bodies or in structures built at the sites. It appeared to me, therefore, that they must have developed ways to discharge or release the vibrations on a regular basis. Strong fields of energy, regardless of their positive or negative qualities, are keenly felt by people open to receiving the charge. The ancients were particularly susceptible to such potency because they were not blocked by intellectual concerns and their lives were governed by the natural flow of the universe. I believe the Mayans of Yucatan built their civilization over strong energy centers and, deeply affected by the power, sought every possible means to discharge and conduct vibrations. The Mayans committed ritual murder and mass decapitation which may have been one effort to release the powerful forces. I observed the violence of their methods on the walls of temples at Chichen Itza, where skulls and crossbones were the prevailing motif. During athletic events and religious rituals, a field of disharmonious vibrations was discharged that I believe still remains at some Mayan sites. At the ball games, players lined up on either side. The first man decapitated his opponent, allowing the blood to gush into two tributaries, one in serpent form and one carved into the stem of a plant with flowers. This violent exchange continued until many players were slaughtered as "appeasement to the Gods."

At the nunnery in Uxmal, three members of our expedition felt the effects of these rites when they almost plunged headlong to the floor of the arena during an evening lecture at the ruins. Audrey Glober, an aerospace executive, told Jean Megrdle and Rita Stricklin that she felt an uncontrollable urge to jump into the quadrangle. The other women, seated far from each other, admitted feeling the same compulsion. Their stories had no rational explanation except that the energy of ancient religious

Wall of skulls at Chichen Itza Yucatan. Do these symbols glorify the mass murders of the ancient Maya used to release the powerful energies?

ceremonies may have been latent in the area. The women may have aligned their consciousness with the negative field, thereby being literally "sucked in" to the space below. Denis Postle, in *Fabric of the Universe*, explains that energy in a magnetic field can be wound up like a spring. Triggered by consciousness, it exerts a powerful effect on the minds of people. The process can be compared to a mechanical toy which is wound up and awaits the release of its lever before it goes into action. The energy at the nunnery may have been in this prereleased state when the collective consciousness of the women caused it to generate potent vibrations.

In times of war throughout history, the urge to kill has

sometimes become contagious. Energy, released through the killing, has regenerated and acted like a forceful magnet, attracting more of its own power. Continuous combat, by its nature, reinforces disharmony. Fields such as the nunnery at Uxmal may have transmitted the death feeling to people who were sensitized to the vibrations. Mayans, like other ancient cultures, may also have realized that such fields of moving energy temporarily disappeared due to the formation of a vacuum, after their discharge. Since displaced vibrations soon returned to the same area through a natural process of recirculation, humans again pick up the negative vibrations into their bodies even though during the period of emptiness, a calm may prevail. It is this peaceful state between periods of strong energy that mankind has always tried, unsuccessfully, to maintain. During the brief periods of peace between wars, in the past 3,000 years, tranquility has been momentarily secured. Yet due to the occurrence of the warlike vibrations, man continues to use fighting to dispel the force. This continuous spiral of negative activity has been self defeating since it has produced a dense physical environment on the planet from which it has become harder and harder to escape, ultimately poisoning the course of civilization.

The Mayans may also have released energy by taking daily enemas. This important ceremony, portrayed in wall paintings, removed toxins from the system and accelerated the flow of vibrations. In Eastern religious cults yogis purified their bodies in the same manner by fasting to develop religious acuity. The Yucatan was a barren limestone shelf where plant and animal life was sparse. The lack of food and the enema ritual may have helped keep the bodies of some Mayans in a perpetual state of higher consciousness. After the cleansing, especially among the priesthood, they received sensory data (just as we now extend our minds into new areas of time and space through the language of science). I believe that the ancients connected with the higher states of consciousness to acquire information which was as valid as our present methods of acquiring data. In the *Center of the Cyclone,* Dr. John Lilly states that our whole planetary atmosphere may be part of an intèlligence network

Male phallic symbols at Uxmal Yucatan Mexico. Absence of male/female polarity of energy among the ancient Maya may have created a negative force field at their sites.

whose code we have failed to break, and whose transmissions are outside our present knowledge. In other words, there could be a universal consciousness which transmits a form of knowledge that we cannot yet understand. Since unrecognized signals have been received from outer space I believe some of these communications may be part of this unknown intelligence. The Mayans like other ancients "tuned in" to such intelligence by releasing the denser energies and probably opened their consciousness to receive the higher centers of information that were naturally available to them.

Mayan priests were probably the exclusive receivers of the knowledge. Before and after divination, they used sweat baths, enemas, and other cleansing techniques. They also practiced self-torture by passing a spiked cord over the tongue. After they achieved purification, they were capable of connecting with higher sources of information, later used to develop a technologically advanced society. Visitors to the sites are fascinated with the pyramids and astronomical observatories which resulted from such knowledge, but the Mayans also had violent means of discharging energy, causing misalignment with the harmony of the universe. What I am implying here is that there are two kinds of experiences at ancient sites. The average visitor simply sees the existing structures and feels awe at such accomplishments. A more sensitive traveller can be deeply affected by feeling the energy centers in the areas. If the force field is heavy, which I believe is the case in some parts of Yucatan, the site may be oppressive. If the energy is lighter and vibrates at a faster rate of speed, there may be a feeling of joy or exhiliaration. Among the Maya, where it appeared that sexual activity was, in part, repressed, the violent methods of discharge created a density still detected by sensitive visitors to the sites.

In addition to Yucatan, powerful force fields in the ancient world were created in the Mediterranean by peoples who conducted elaborate sexual rites combined with religious ceremony. The Mediterranean sites, however, were infinitely more positive than the Mexican ones simply because the discharge mechanisms relied more on the positive sexual approach. The experiments of Dr. Wilhelm Reich suggest that sexual energy, called the orgone, produce a vibrational field of great intensity. In Reich's experiments a large box known as an accumulator was used. The box was made of galvanized zinc, with layers of cotton and screens interspersed to make a thick outside covering. Inside the box was a fibrous board which Reich believed absorbed the orgone essence. The box also contained a magnifying glass. Reich placed subjects in the box and observed their auras or energy fields as they stayed inside. He also taught his subjects how to breathe properly and how to maximize the effects of the box. Reich inferred from his experiments that there was a change in the rate of vibration given off by the subject as measured by the colors he observed in the aura of the participant. At the time he performed his research, he was severely criticized by scientists but, more recently, books on Reichian therapy have been widely disseminated and the value of his experimentation is beginning to be recognized. In contrast to Mediterranean people the later Mayans of Yucatan may have feared the sexual power of orgone and invented human sacrifice as a safer method of discharging the energies. This philosophy contradicted the earlier stages of Mayan civilization where sexuality probably was revered as evidenced by the uplifting energy in the archaeological centers of Guatemala's earlier Mayan sites.

On the island of Isla Mujeres, in the Mexican state of Quintana Roo, a female cult emphasized the unity of the sun and moon, or the blending of masculine and feminine forces. This natural philosophy about the regeneration of the flow was positive and uplifting, as in the cultures of Egypt and Babylonia, which also stressed the polarity principle. According to Alan Watts, sexual activity has always been necessary for the soul as well as the body. In his book *In My Own Way—An Autobiography*, he states that sex provides an ecstasy and a timelessness that eliminates the separation that many people feel from their environment. Some ancient people understood this principle and respected sexual exchange as the most positive form of energy discharge.

Most Eastern religions had an opposing view of the role of sexuality in the purification of consciousness. They maintained that of the seven body *chakras* (energy centers), the sexual *chakra* was the lowest form of release and the crown (brain) *chakra* was the highest. In the later stages of the ancient Mediterranean world this taboo also prevailed and there was a decline in the use of sexual activity to release energy. Problems of overpopulation and disease, as well as the fear of the force itself, caused a repression that eventually entered the Judeo-Christian heritage and has prevailed up to our own time.

The revolution in consciousness of the past several decades was connected to the liberalization of attitudes toward sex. Most young people came to realize that there was a relationship between heightened perception and sexual flow. Early attempts at liberalizing sexual practices alarmed educators and religious leaders but the palliative effects of sexual activity were eventually accepted by some more forward thinking members of the establishment. The urge toward freedom in sexuality among the youth subculture actually emulated the practices of the ancients in connection with energy discharge and natural balance. Unfortunately, the search for higher states of awareness was also connected with drugs which suppressed the flow of energy and worked in opposition to the natural principles. Eventually, the first stage of the youth upheaval was transformed to a later, more positive, development where meditation, diet, and safer techniques to raise the vibrational flow were emphasized. Thus young people naturally acquired the concepts of ancient energy as they sought to reestablish the same principle of oneness that was known to earlier cultures. Their search for answers in a transitional society created a renewed interest in ancient wisdom. As the cycle started over again, it confirmed a belief held by the Maya, who contended that everything was continuous including the eternal cycle of life and death.

Ancient energy flourished wherever diverse cultures continuously occupied a site. The Loltun caves, near Oxkutzcab in Yucatan, were inhabited for centuries before the Maya culture existed, and were submerged by flood several times in the history of the peninsula. The natural grottos were explored in 1965 by

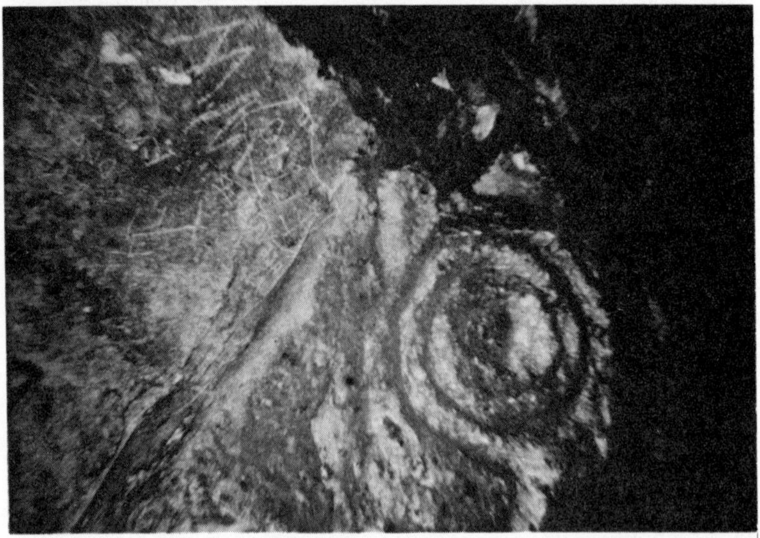

Double spiral petroglyphs in the Loltun Caves of the Yucatan are identical to markings found in Ireland, Malta, and other Mediterranean sites. Could they be transmitters and receivers of the energy flow?

Dr. J. Manson Valentine, who photographed spiral petroglyphs which he believed to be more than 15,000 years old. The 28-chamber cavern, still under excavation, has linear writing, phallic symbols, sacrificial altars, and red ochre paintings. Despite the tortuous passages and steep cliffs, Loltun transmitted a highly intense and positive vibration. The facilities were primitive but I welcomed a chance to be one of the first to explore the site.

On my first expedition to Loltun in 1978 our party was only halfway through the caverns when the electricity failed, leaving us in total darkness. I stumbled along the rocky floor after the guide who beckoned me to accompany him to a death chamber where black hands were painted on the cave walls. The cavern was a ceremonial center which the Mayans venerated by tracing around the hands of the dead with black paint on the upper portion of the room. I was fascinated by the site and cautiously stepped down a pile of rocks to photograph the symbols while my friends extended their lanterns. As I turned to walk back up, I had a sickening, light-headed feeling similar to that I experienced at the lower tunnels of Abydos. There was adequate air in the caves, so I knew that a vibration more intense and concentrated than the energy in the open air arenas of Uxmal and Chichen Itza must have thrown me off balance. Perhaps the discharge of energy through the creation of the hands of death had obstructed the natural balance of the chamber which had not only affected by own perception but also that of Bill Davis, who also had some temporary dizziness and confusion. Bill, a geologist from Nevada, had a purely scientific interest in the cave, so he was startled by his reaction to the energies in the death chamber which he had not anticipated.

In *Mystery of the Ancients*, Craig and Eric Umland claim that the interior of Loltun was more powerful than the Cheops pyramid in Egypt. Other hill caves of Yucatan were equally intense. As we approached the entrance to the Caves of Balanchine, I experienced a feeling of suffocation and a strong intuition to return to the hotel. Later my friends admitted that the investigations had been arduous even for veteran cave explorers. Several people became physically ill from the heat and narrowness of the tunnels as well as from other factors. Others claimed that the concentrated energy inside caused them to feel panic and some transition of their consciousness. Since perhaps sacrificial rites had been practiced in the caves, they may have created an imbalance in the force field similar to that in the death chamber at Loltun.

The discharge of energy at ancient sites did not affect everyone in the same way, possibly because different people received the vibrations at different levels of consciousness. When two or more people operate at the same "wavelength," as did the three women at the nunnery in Uxmal, they may share a common interaction with the forces. While I suffered through the death chamber at Loltun, Meg Downs passed by without incident. Later she told me that at another site blinding headaches and physical problems forced her to shorten the trip. Meg, a businesswoman, had felt a general uneasiness when she arrived at Cairo, Egypt. The feeling grew more intense at every site she explored. Yet, suddenly, five miles out of the Alexandria harbor, all discomfort had disappeared. My experience in Egypt had been totally different. Though Meg and I both sensed the heavy force fields of the sites, our reactions were not the same, since we conducted the energy at different levels of awareness. I also believe apprehension, jet lag, and physical problems of people who travel may be caused by an abrupt shift into a new center of vibration, especially at Mayan and Egyptian sites where fields of discharged energy are heavily concentrated. Pyramids in these cultures were used for initiation and divination through sacrifice. They were constructed over high energy centers on the earth's crust that maximized the flow. The strongest reactions to ancient energy often occurred near such pyramids. At Palenque, in southern Mexico, several of our expedition members felt disoriented as they approached the Temple of the Inscriptions. Gradually, they adjusted to the higher vibrations, which stabilized closer to the center of the energy source, in the burial chamber beneath the pyramid's base. At Chichen Itza, the dense vibrations at the pyramid El Castillo may have resulted from the dense energy released by women and children who were thrown into nearby cenotes (natural wells). Today the environment still transmitts this disharmony to the sensitive investigator.

Not all Mayan sites were the same. I sensed that the ancient Guatemalan centers were totally different from those in Yucatan. There was an immediate feeling of peacefulness at Chichicastenango, the mountain city of the Quiche Maya, whose laws of energy originated in mythology. The Mayan Bible, Popol Vuh, reinforced the principles of oneness. According to tradition, if a piece of white corn and a piece of yellow corn were left alone on a patio, they would fertilize of their own accord, since a single essence pervaded the cosmos. Modern descendants of the Maya practiced unity in their ceremonies by burning colored candles and rose petals to simulate a balanced male-female vibration. They dispelled illness by allowing energy from the ceremonies to enter the body of the patient, which was similar to the practices of the Navaho and Yaqui medicine men. Also, incense was burned around the steps of the church to complete the circle of protection. At Chichicastenango, one sensed nonviolence, balance, and harmony similar to the feeling at ancient Mediterranean sites.

Shortly after visiting the Guatemalan ruins, I tried to understand the difference between the force fields of Yucatan and Chichicastenango, both Mayan centers which transmitted totally different vibrations. I was enlivened by the Quiche Maya and depleted in the sacrificial sites of their ancestors. My arms and legs stiffened at Chichen Itza and Uxmal, and I was unusually sleepy there. Later, when my fingers and toes became numb, I sought a medical explanation for my body's reaction to ancient energy. Failing to find any conclusive physical cause, I reasoned that an energy mass incompatible with my own had entered my endocrine system, nerves, and muscles via the brain and neural pathways. When areas of density placed additional pressure on these organs, normal sensation was blocked. Where vibrations were less dense, buoyancy resulted. Thus, the Quiche Maya probably raised the vibratory rate and obtained beneficial physical results.

These ancient concepts could have significance for our own life. Perhaps there is a way to reverse the present trend of controlling our environment through technology, by allowing the natural unfolding of healing, creativity, and harmony. The possibility of curing the physically or mentally ill through the realignment of body vibrations could radically change the way we view traditional medical practice. Environmental scientists might become concerned about the inner environment and study the relationship between consciousness and planetary survival. Possibly we could develop a stable partnership with nature for the first time in five hundred years. Ancient energy may be the universal principle we are seeking to restore a sense of individual well-being and peaceful interaction among groups.

Through a process of trial and error, I learned to discharge heavy vibrations at archaeological sites using some of the traditional methods of the ancients. These included a daily ritual of burning candles, meditation, and body cleansing and exercise. The two-hour process of mental, physical, and spiritual purification gave me unbounded energy. When the vibrations were very dense, I conducted the practices several times each day. Liquor and cigarettes were taboo, since they repressed energy rather than channeled it. When I was questioned about my bizarre habits, I reminded my critics that anthropologists often live among primitive cultures to learn their ways. After I had immersed myself in the pagan rites, I had a better understanding of how the ancients transferred the energy flow. Friends marveled at my incredible stamina. "Ancient energy," I told them. "You can acquire it too." I explained to them my method of awakening before 6 A.M. each morning and described the ritual I followed. First I lit a candle, turned on some soft music, and said a short prayer for divine protection during the day ahead. Then I meditated and engaged in some brief physical activity like swimming or jogging. Finally, I prepared a breakfast of freshly squeezed orange juice, eggs, dry wheat toast and coffee which I always ate propped up in bed while reading the morning paper. The entire revitalizing process took less than an hour, which left me another hour to work on my writing before leaving to my job at the school district. I never abandoned this routine on vacations or holidays and I also performed as many of these activities during the day as time allowed. I also taught my methods to people who accompanied me to the sites although not everyone was prepared for the discipline required to maintain the procedure, which only works if practiced on a regular basis.

Ancient people such as the Inca and the Maya also performed daily ceremonies using prayer and cleansing in their rituals. Incorporated into such activities was a strong element of faith. I reaffirmed the importance of strong belief in making the energies work for me when I conducted explorations in the mountainous areas of South America. I was enroute to the Caves of Tayos in southeastern Ecuador where I planned to investigate the work of Eric Van Daniken, who claims he saw 20,000-year-old gold plates inscribed with primitive writing inside the caves. During the first stage of my trip I flew from Quito to Cuenca to visit Father Carlo Crespi, who showed me his amazing collection of ancient artifacts and gave me information about Van Daniken. Since I had to go to Tayos via helicopter into the jungles, I opted instead for a visit to Ingapirca, an Inca fortress fourteen thousand feet up in the Andes. On the way to the site, oxygen starvation caused me to feel very dizzy. My problem was so serious that I told the driver to act quickly since I was sure I was going to faint. He answered in Spanish, saying simply "Pray," and pointing to a statue of Jesus on the dashboard of the car. "Pray?" I thought. "Here I am half dead and this guy says 'Pray.'" I slumped down into the seat as we wound around the three-foot-wide highway that led to Ingapirca while the driver assured me that prayer would restore my equilibrium. Since it was impossible to turn back, I began some silent meditation, imploring the great gods to take me out of my misery. I had visions of dying in the Andes or ending up in a cannabilistic ritual like the survivors of the air crash in the book *Alive*. But, to my surprise, thirty minutes later we arrived at Ingapirca and my altitude sickness was gone. I jumped out of the car to investigate the ruins, explaining to the driver that I could not imagine how the situation had reversed itself. As we contemplated the fortress, he told me that the ancients used prayer in response to any problem, since all human survival was assured by direct communication with the gods. The Incas believed that they were favored because they lived close to the source of universal knowledge in the high altitudes of the Andes. Like the Maya of Guatemala, Inca ceremonies involved the burning of incense, the use of mild narcotics such as coco leaves, and the chanting of powerful incantations to raise the vibrational field. Some of these sounds are still made by present-day Incas, whose ceremonies recalled the earlier rites. Hymns of cultures are often carried through the ages. The songs of the Rig-Veda in India were recreated when the priests documented the exact positions of lips and tongue as the sounds were made. The Incas developed whistles that produced monotonal vibrations strong enough to throw spectators into an altered state of awareness. Primitive people retained their own sacred vibrations that may have harmonized with a universal intelligence, still imperceptible to our present society.

From my observations, ancient people who lived in high altitudes had a common cultural heritage. The Incas, like the Tibetans, built great pyramids and temples with ease, using blocks that weighed dozens of tons. Tiahuanaco in Bolivia, Machu Picchu in Peru, and Ingapirca in Ecuador were constructed with a technology different from our own. I believe that sacred knowledge may have been more easily accessible to the ancients in the less dense mountain environment where extensive human sacrifice and violence were not required to clear perception and raise consciousness. For example, Moses went to Mt. Sinai to "receive" the Ten Commandments, the Greek gods discussed man's fate on Mt. Olympus, and caches of sacred knowledge have been reported in such remote areas as Mt. Shasta and Tibet. The people of antiquity who lived in the mountains may have known the principles of weightlessness. It is my speculation that they may have removed the specific density of blocks and transported these building materials to the sites where they lifted them into place. The universal intelligence by which ancient people invented calendars, engineered the megaliths, and performed precise mathematical and astronomical calculations is available to us in the form of ancient energy lying dormant at archaeological sites around the world. We could greatly expand our potential for new knowledge by tapping into the vibrations and translating the information received into ideas that harmonize with existing scientific data. As an example, let us say we are faced with a practical problem of how to eliminate smog in the Los Angeles basin. Practical, rational approaches to the issue have been tried but have failed to accomplish the desired results. I believe it could be beneficial for a group of civic leaders, engineers, and concerned citizens to retreat into a meditative state to develop a state of consciousness by which they could discern some natural principles applicable to the problem. I feel reasonably confident that with proper motivation and dedication to such a new approach, creative solutions might emerge which, when combined with present systems, might help reduce the problem of smog.

I believe that ancient energy has characteristics that are different from all other substances in the universe. I also strongly suspect that it has always existed, cannot be observed, is experienced by humans, and is not easily replicated. This elusiveness causes many people in our present culture to deny the existence of the phenomenon. Even those who acknowledge its physical effects cannot quantify or describe it. Frequently, the energy appears to work in reverse, and there are no known laws governing its behavior, except for random examples drawn from quantum physics. Only in recent times, since 1970, have a few scientists become aware of its importance for scientific inquiry. Experiments with auras, or energy fields, by Dr. Thelma Moss at UCLA, and the influencing of motion through the power of the mind by Luisa Rhine at Duke University, have made a great contribution to the understanding of how such energy functions.

Man's scientific accomplishments are unsurpassed, but problems in the healing arts, environmental sciences, and global communication still lack solutions. The investigation of ancient energy may lead to the development of new methods of restoring the natural balance known to primitive man. A revolution in the way we view our universe is mandatory for such research. It will not be easy to bring about, since researchers are primarily trained to accept a physical universe with cause and effect relationships. Even if scientists were to include a phenomenon such as ancient energy within the scope of their research programs, the nature of the measuring instruments currently available is inadequate to deal with nonmaterial entities. Finally, human resistance may be the most difficult barrier in establishing ancient energy as a field for research. Since the Middle Ages, most significant scientific issues have not required a major consciousness change on the part of the researcher, even though insight and creativity were always important. If ancient energy is to be dealt with in the laboratories and universities, the investigator must change his or her entire structure of reality, something most people find uncomfortable, if not impossible to accomplish. Thus, discovering ancient energy depends on a restructuring of scientific intellect and a realignment of global consciousness as well.

There are certain concerns in connection with such research. Controls are essential for the maintenance and use of the force, just as the creation of laws has been necessary for the wise use of nuclear power. Ancient energy, carefully managed, could reverse the pattern of violence and warfare which has been a part of Western civilization for milleniums. Few people have been trained to manage vibrations in order to exert a positive influence on their own lives. A psychiatric nurse told me that techniques of ancient energy may be used with mentally ill patients whose erratic behavior appears to be a problem of perception. Hysteria could be seen as a blockage of energy and schizophrenia as an imbalance. During hysteria when the patient goes out of control by screaming, crying, or performing violent physical acts, he releases energy which eventually results in a period of temporary calm returning eventually to the hysterical state. In schizophrenia, it is possible that too many frequencies of energy may course through human consciousness, causing violent swings in temperament. Patients with either one of these two diseases may lack the controls to keep the energies in harness or have not learned appropriate methods of discharge. In my opinion, administering medication to such patients cannot resolve the problem which might respond better to meditation, athletics, and other more natural energy outlets. Ancient societies forestalled madness by providing other means for energy discharge. Roman games and Greek athletics were acceptable ways of recycling vibrations. At Delphi, Ephesis, and other oracular sites, sexual exchange and religious ritual were combined to produce a frenzy of activity which also unblocked the flow. Nearing Christian times, the older, more positive approaches used to discharge the forces were abandoned in favor of negative means such as isolation, starvation, and self torture, especially among the early Christian ascetics who reinforced the idea that the "good life" came from suffering. This reversal in technique may have come about when the pendulum moved violently from decadent Rome, with its emphasis on pleasure, to the simple religious life espoused by the Christians. Suffering became a more spiritual option for relief from what were then called "evil forces." These ideas were contrary to the natural goodness of humanity since I believe that a joyous (but not decadent) state is the ideal for which humankind must continually strive. Despite the impact of heavy religiosity, society has been out of balance with the flow of energy for the past 2,000 years.

Like the Egyptians and the Mayans, most ancient societies respected and incorporated energy into every aspect of their lives. Loose, simple clothing reflected and expressed natural rhythm. Loincloths left the chest exposed and the female breast was frequently bared, as in Crete. Materials were draped about the body to enhance the natural flow. Kaisek Wong, a San Francisco designer, recreates primitive costumes based on ideas that appeared to him in clairvoyant dreams. His clothing encourages flow in all of its many dimensions. Other industries have recently returned to techniques used in early times. Architects study ancient building forms and pattern entire residential and business complexes after structures created by ancient architects.

Greek and Roman homes were designed to allow natural light and varied colors to enter. Areas were exposed to the rays of the sun and the moon. Alignment with other planetary phenomena was also part of the original design. I suspect that the Mayans slept in hammocks to ensure receptivity to the earth's vibrations since personal inquiries made in the Yucatan led me to believe that they are still used for the same purpose. I also think our contemporary return to water beds was based on similar principles, since water beds cause a natural flow of body movement throughout the period of rest. Rocking chairs, an old standby, may also have accomplished a similar purpose, that of keeping the energies active. High speed rides at amusement parks and daredevil feats of all kinds may be popular because they keep the energies in constant motion. During the 1960's, youth returned to the long hair worn in ancient times, emphasized long flowing dresses, coats, and jackets, grew beards, and, in all ways, increased the possibilities of body movement which aided in discharging the energy flow. Such thoughts are alien to our Western tradition but may be linked to the concepts of ancient energy. By reflecting on some of these ideas, we may come close to understanding the nature of the universal essence so that we become one with the world rather than at odds with it.

I believe that all people are affected by energies although they perceive and deal with them in different ways. Visitors to ancient ruins generally have positive experiences, but they may endure headaches and other discomforts until they have adjusted to the environment of the site. Such random encounters can be developed and encouraged so that the force works for rather than against the individual. The beneficial effects of ancient energy are available, but they must be channeled through consciousness and released by means of the positive techniques employed by the Egyptians, Mayans, and ancient Mediterranean cultures. Such practices include meditation, the use of color and form, daily exercises to expand consciousness, and a steady awareness of the forces around us. These ideas will bring us closer to the natural understanding of the universe known to most ancient people.

7. Ancient Energy and Atlantis Research

On February 13, 1973, a news conference was held at the Los Angeles Press Club announcing plans for an underwater search for Atlantis off the coast of Cádiz, Spain. Dr. Julian Nava, professor of history at California State University Northridge and a member of the Los Angeles City Board of Education, had consented to codirect a scientific expedition to locate remains of the fabled continent. Divers, professors, and researchers were invited to take part, although the trip was open to the public at large. Halfway through the meeting, Harvard-trained Dr. Nava announced that intuitive techniques would be employed in the search and that I was the one who would use them. He indicated there were "other ways of knowing," and gave examples of myths that had been verified by archaeologists based on their hunches. Citing Troy, Mycenae, and Cnossos, Dr. Nava explained that the archaeologists who made these discoveries used methods considered questionable by scientists of their time.

At the end of the conference, newsmen asked me about the intuitive component of the expedition. They were astonished that academics such as Dr. Nava and myself would openly advocate methods that were disdained by the "ivory-tower set." I had little to say to the press. Julian had not warned me that he was planning to make the announcement, even though he knew I had an interest in energy applied to the process of anthropological investigation. His statement had taken me completely by surprise. Two weeks earlier, on a research trip to Cádiz, he had vehemently denied the existence of such forces and urged me not to state my interest in the phenomena to the press. My publicity agent had given the same admonition. "It will ruin you," she said. In the period following the conference, Dr. Nava continued to acknowledge the strong possibility that Atlantis could exist and that the intuitive method was one way of approaching the problem. However, with so many other academics I had observed, their first encounters with ancient energy frequently left them feeling the forces were unmanageable. A premonition, dream, or strong exposure to the denser vibrations often caused them to close their minds completely to the possibility that they could be affected by something unproven by laboratory experiment. It was easier to shut down the perceptual doors than to be involved in such research. Julian was barely informed about the subject of energy in spring of 1973 and yet he was an early proponent, with an open mind toward the subject. Later, he became a partisan of such methods and lent his support to the Psi Search Exposition that appeared nationwide from 1974-1977.

Julian was unique in his acceptance of new approaches to problems in the social sciences. Most members of the academic world, even when agreeing to such possibilities, ignore actual problems of research. I was one of the few investigators who disagreed with the established view of man's origins and the methods used in field research. From my first inquiries into the problem of Atlantis, I realized that a new way to tackle the question was essential. When a continent disappears more than 10,000 years ago, the evidence for its existence is scarce; there are few, if any, credible books on the subject. In the case of Atlantis there has been little field work except for the intuitive writings of Lewis Spence and the travels of psychic Madame Helene Blavatsky. To demonstrate the existence of Atlantis, I searched for artifacts or books in the United States but the only literature I could find was mythology, the writings of Plato, and dozens of psychic accounts. No academicians cared to comment on Atlantis and there was no possibility of interesting the intellectual community in supporting such field research.

Failing in my efforts to support the shaky hypothesis that Atlantis had really existed, I travelled to Europe, the Mediterranean, Central and South America in search of inscriptions, artifacts, and printed material which might establish Atlantis as a scientific reality. I collected rocks, photographed sites, taped interviews, and combed the countryside searching for clues. Most of my time was spent in the south of Spain, which matched Plato's vague description about Atlantis, as situated near the Gates of Hercules (Gibraltar and Tangier). I had considered other locations for the possible existence of Atlantis like Greece or Ireland. Yet, my hunches seemed to point to Spain in the southwest corner of the Iberian Peninsula at Cádiz, where artifacts, interviews with divers, and mythological accounts of the lost continent were most prevalent. It was difficult, however, to pinpoint the location because I was frustrated by piles of data that never came together in any logical form. I felt no matter how much data I accumulated, there was a missing link needed to move my ideas from speculation into reality. Leakey had searched for a key to explain man's earliest origins. Many scientists waited for that moment of "creative insight" when their theories were finally explainable through good supporting data.

Most of the books on Atlantis are esoteric accounts written by people who fabricated elaborate stories of the lost continent, replete with devastating crystals, light that resembled laser beams, submarines, UFOs, and all manner of Jules Verne constructs. I was dismayed at the lack of objectivity on the subject, but I understood why people continued to write without scientific verification. Interest in Atlantis had persisted for thousands of years, even without scientific proof, and the public demanded books on the subject. Serious Atlantis scholars gave up their search early in the game because of academic ridicule, the dearth of evidence, and the lack of available money. Still, popular feeling on the subject is so strong that twenty thousand books and materials are believed to have been written on the subject of Atlantis. All have been based on intuitive memory, hunches, or elaborations of Plato's story. Such memory, drawn from the imagination of creative writers, was formerly thought to be purely fictitious narrative. However, so many writers have retold the same stories about the culture of Atlantis, its origins, and its destruction, that I believe it is possible that such intuition is based on the reception of information coming from ancient energies at higher vibrations received by these people. I strongly suspect, therefore, that a memory exists about Atlantis which is not a random occurrence. At one point, I changed my home telephone to an unlisted number when I began to receive as many as several dozen calls per day about Atlantis and its civilization from people who had dreams or visions about its existence.

Plato's account of Atlantis takes over a dozen pages in his Dialogues "Timaeus" and "Critias" which describe the location of the continent, its society, and dozens of other details, including the date of destruction, 9600 B.C., which was recently verified by marine geologist Dr. Cesare Emiliani writing in the September 26, 1975, edition of *Science* magazine. Dr. Emiliani measured water levels around the Great Lakes area in order to build his theory that a worldwide flood occurred at the time Plato reported the sinking of Atlantis. The myriad Atlantis stories often reflect Plato's account, but Dr. Emiliani's data was the first scientific support. I suspect that both intuitive writings and physical reports should be considered as information in piecing together the riddle of the lost continent.

Fortunately, Europe's bookstores were better supplied with detailed accounts of the geography, geology, and history of the lost continent. I accumulated a vast collection of volumes, but I continued to rely heavily on Plato's account. He claims to have learned about Atlantis from Egyptian priests at Sais near Alexandria. Since Egypt had been a focal point for my research, I wondered about the esoteric accounts of Atlantis written by Edgar Cayce who claimed that refugees from the continent arrived in Egypt about 10,000 B.C. The timing also corresponded with the appearance of the advanced Mouillian culture in Algeria, believed by some investigators to be created by Atlantean survivors who fled to North Africa after the flood. Dr. Emiliani's work demonstrated how accurate Plato was in his dating, and it was more than coincidence that so many other events seem to come together around 10,000 B.C. Yet Plato received his story in the fifth century B.C., almost ten thousand years after the destruction occurred. Perhaps in addition to the religious testimony he received in Egypt, he may have gained intuitive information from the ancient force fields of the Mediterranean, including Spain, Italy, and Greece. His accounts of the Atlantean use of land sleds (still used in Madeira), quarries of red, white, and black rocks (abundant in Andalusia), and circular cities (spiralled like the triple petroglyphs of the Canaries, Crete, and Newgrange) are very skillful descriptions and hardly guesswork or coincidence.

For years I tried to verify Plato's location for Atlantis, and found the most promising site was Cádiz, where artifacts, fossils, petroglyphs, and specimens which might be related to the lost continent exist. In addition to my own fieldwork, I reexamined other documents previously thought to yield only mystical data such as the Old Testament, Greek mythology, and the Cabalist writings of the Hebrews. They also provided clues to the growing bank of information supporting the existence of Atlantis. In Cádiz I interviewed taxi drivers, waiters, people in the street, and professors at the university. Everyone had something to say on the subject, although most had never read or studied about the problem. They simply *knew* through intuition where to find information. They spoke of Atlantean writing, fossil evidence of the flood, burials, and mounds, which they believed were related to the object of my search. Professors at neighboring colleges and Spanish admirals gave me information, maps and directions that kept me exploring hours each day.

The attitude toward Atlantis in Spain, and the widespread folk legend among the vast majority of Spaniards contributed to a deeper awareness of the subject, in great contrast to the paucity of information in the United States. Skepticism and disdain characterized the attitude of many of my colleagues when I mentioned the subject. Not only did they deny the existence of Atlantis, but also they refused to examine my slides and specimens. The nagging question of why so many people in Spain were conversant on the subject brought me to an examination of the theories of Jung, who claims that symbols filled with information about the past come into reality for some people through the collective unconscious. I believed that people who were immersed in the strong energy fields of Spain gained information in this manner because memories about Atlantis, buried in the unconscious, were remembered more easily in the higher vibrations of Iberia. In other words, the energies served as a channel to open the unconscious mind.

The closer I came to Gibraltar, the more useful was the information I obtained. Physicist Robert Oppenheimer explains such behavior by indicating that there are two ways of thinking, "the way of time and history and the way of eternity and timelessness." He claims that both are part of peoples' efforts to comprehend the world in which they live. Each way supplements the other and neither tells the whole story. The way of timelessness is a channel to the ancient cultures that have disappeared. Collecting and documenting information from people who "remember" the past has definite value for Atlantis research where physical evidence is sketchy and circumstantial.

Jung does not explain the nature of the unconscious stream that surfaces in the form of information, but I sensed that it is especially potent in certain cultural areas. As a result of my travels to Mexico, South America and Spain, as well as my public school work in bilingual education, I observed that Latin people operated at an affective level of consciousness far more than non-Latins. This factor, I believe, helped me to gain more information about Atlantis in Latin-speaking countries since the people appeared to be more open to such concepts and more literature existed in bookstores and universities related to the subject. Part of the openness to topics like Atlantis may be encouraged by the "feeling" component associated with the Spanish and Italian languages, or a vibration which accompanies the basic language structure and helps the perceptual centers that receive the higher energies stay open. Shutting down of the brain's capacity to handle these vibrations may come when people from a Latin-speaking country are transferred abruptly to the United States, a phenomena I noted with some frequency in the classroom. Students appeared to suffer a kind of culture shock as a result of the swift change from one language type to another. As I attempted to teach language transference from

English to Spanish by opening the perceptual centers, I believe I helped students keep both hemispheres of the brain open which ultimately led to their ability to acquire information via the same sensory pathways as those used in the conduction of ancient energy.

My task as I perceived it was to integrate the individuals' intuitive or memory data bank with the information from physical artifacts and specimens in order to make sense out of both. People, it seemed to me, were as important in solving the problem as rocks, inscriptions, or artifacts. Yet I was faced with the dismal prospect that anthropologists would never accept human testimony as proof in the scientific sense. I searched for a statistical instrument that could measure the kind of information I was obtaining, but I returned again to Interrater Reliability, a collection of baseline data that plotted people's opinions relative to a common problem. The other method was observation, and documentation of data in journals. Neither approach was satisfactory, although I had attempted in my doctoral dissertation to measure the amount of intuition and affective data used by anthropologists and historians in their research. I had polled several hundred professionals nationwide. The results indicated that they used a subjective or an intuitive approach to laboratory and field investigations at least thirty-five percent of the time. The report suggested that if professors were acquiring at least part of their data in nontraditional ways, then perhaps a new methodology could be developed for use in anthropological research by using the expertise of academicians who were open to such approaches. Yet, in spite of the data I had secured, I remained hesitant about the use of such techniques in connection with the study of Atlantis. The subject itself was anathema to most scholars and the use of such an avant-garde investigative approach would be totally unacceptable in establishing positive proof for the existence of the lost continent. Thus I tried to use a variety of methods to avoid having my work dismissed as the research of a mystic, despite my degrees, credentials, and high standing in the educational community. I knew that underwater research was ultimately the only method that would document the existence of Atlantis. Thus, the focal point of the search for a

The author and diver Francisco "Paco" Salazar examining a Phoenician vase taken from the offshore waters near Cádiz Spain.

Paco continued to dive since he was a Spaniard and in high favor with the naval officials.

Our divers were affected by the problem of rapidly changing levels of consciousness which from time to time I had also noticed in Julian and the other professors. For example, on several occasions, Bill could clearly describe the remains of the city at Zahara, while on others his memory was less accurate. The murkiness of the Atlantic Ocean was part of the problem, since sand and silt quickly cover undersea discoveries in that area. But I believe another factor was in operation. Levels of awareness differ from person to person and vary at times within the same person. The "now you see it, now you don't" phenomenon was a common occurrence among individuals and groups in sites of concentrated ancient energy. One of our most capable divers, skilled at using the new methods in his research, was Gary Varney, who acknowledged that changing levels of consciousness may influence underwater research. Gary did not stay in Cádiz long enough to make discoveries in Spain, but he discovered many artifacts using an intuitive approach to his investigations during 1974 at Bimini, as reported by Dr. David Zink in *The Stones of Atlantis*.

In addition to the undersea observations, we worked directly with students who were taking regular classes at the Conservatorio, a three-story study center donated for our research by the Spanish government. The course, "Common Cultural Origins of the Ancient Mediterranean World," was accredited for six units by California's Pepperdine University School of Continuing Education. The group was comprised of students and teachers along with laymen, housewives, and esoterists who had come together because of their common interest in Atlantis. I had trained the students in research techniques before we left Los Angeles, so they were prepared to examine all kinds of information regardless of how the data was secured.

One of the instructors, Gail Cayce, was the granddaughter of the famous psychic Edgar Cayce, the Virginia clairvoyant who gave accurate medical diagnosises to thousands of people without direct access to the patients. Cayce often made his prognoses on the telephone, over thousands of miles to people he had never seen or spoken to. One of Cayce's archaeological predictions had been supported in 1968 by Dr. J. Manson Valentine, a Ph.D. holder from Yale University who claimed to have discovered Atlantean temple blocks at Bimini described in Cayce readings. He photographed the blocks and then submitted them for Carbon 14 laboratory tests which produced a date close to 9600 B.C. Fortunately, Dr. Valentine was open to my ideas about ancient energy. But he was careful to advise me about the need for protection when exploring in areas with strong fields of force. At discussions in his Florida home, he cited cases of explorers who had gone to South America, tampered with the energies at ancient sites, and died or disappeared. He also believed in the reality of UFOs and underwater pyramids, which put the academic institutions to which he was attached in an uncomfortable position due to the unpopularity of Atlantis as a scientific study worth of consideration by the university. Yet, in spite of his nontraditional beliefs, no one could deny that he was a hard-core scientist who found empirical evidence to support the Cayce predictions using legitimate underwater archaeological techniques.

The accuracy of some of the Cayce readings is indisputable, but I could not accept all of his predictions without question. His correct identificaton of the Bimini complex was important evidence that the method worked. However, the Mediterranean was another issue. None of Cayce's statements about Egypt, Spain, and France has been proven. In my opinion, he had been too far away from the force field of these sites to achieve any degree of accuracy, despite claims that using a dowsing rod over a book to pinpoint information in faraway locations is effective. Based on my experiences, I believe that the investigator has a better chance to make finds when he or she is submerged in the energy field of the site. The process of discovery, using an ancient energy approach, depends on the intensity of the vibrations, which may diminish the farther away the subject is from the field of force. For example, Edgar Cayce called the last piece of Atlantis "Poseidia," referring to the sea. Most likely the name should have been assigned to the piece of Atlantis near Spain, described by Plato, which is closer to Greece. I believe that Cayce's predictions were correct but his geography was frequently in error due to the distance between his home base and the sites he referred to in his readings.

Our research program in Spain also involved the collection of intuitive data from students, a difficult process even for Dr. Michael Hughes, who assisted us in that effort. Michael had learned techniques from Dr. John Lilly at Esalen Institute and Dr. Brugh Joy at the Center for the Healing Arts in Los Angeles. He had studied consciousness raising, and was interested in defining the essence of spirituality and the memory of times past. At informal sessions in the evenings, apart from the regular curriculum, he gave "readings" to the group about Atlantis. These meetings, which were optional activities for students, forged a unity of ideas within the group and provided an alternative means of gaining information on the subject. I believe that the collective sharing of intuitive ideas is valuable for research when other sources of information are not available.

Most of Michael's activities were held with individual students and conducted privately. Generally he sat across the table from the subject and placed himself in semi-trance while still conscious of everything around him. Then he reported different symbols which he was able to visualize from his subconscious. The student wrote or drew his messages on paper during the reading. Later Michael and the subject created a story from the symbols and tried to evaluate the significance of the data. Many of the readings done in Spain resulted in the same information. There was a consistent memory of Atlantis, primarily in the last days before the destruction, as described by Plato.

The teaching of Edgar Cayce readings on Atlantis by Gail Cayce, and the work of Michael Hughes caused serious concern among the strict scientists who had agreed to work with our group. They insisted that scientific objectivity was the only valid approach, and viewed these techniques as occult or mystical, similar to the practice of alchemy in the Middle Ages. I could not reassure them that such methods had academic validity since the investigation of ancient energy is a new paradigm that is neither scientific nor psychic but rather an examination of how particles of matter behave in the quantum state and the effects they have on human consciousness and learning.

Unfortunately, our research was stopped after only four weeks by mysterious elements in the United States and the Spanish governments, which to this day are not clearly understood. We were branded publicly by the United States press, who questioned the validity of our methods, but we were encouraged by the supportive attitude of the European press which was objective and sympathetic in its coverage. The cause of such restrictive action was based on the existence of U.S. nuclear submarine launching platforms placed underwater off the coast of Spain as well as a coalition of political, intellectual, and religious opponents referred to in my book, *The Atlantis Conspiracy*. The Spanish government had issued official permits for diving through the Department of Belles Artes and provided our students and staff with a three-story public building as a classroom. We had national and international publicity for the search in newspapers, magazines, radio, and television. Pepperdine University accredited the course and Senators Tunney and Cranston of California provided letters of support. Although we never understood exactly why we were stopped, it was obvious to us that the military problem was the greatest obstacle. In view of the harassment, little physical proof of Atlantis was secured in 1973 despite Paco's best efforts to produce it. We had a few photos, the testimony of the divers, and Paco's sketches. This fragmentary data was inconclusive so it was understandable that people continued to doubt the existence of Atlantis. I was not one of them. Evidence from my own research was mounting, and I had to continue the search. I knew that time and effort often result in proving even the most far out theories. The long awaited physical evidence of the lost continent was becoming more of a reality as I continued to collect bits and pieces of data. The obstacles I had faced only caused me to pursue the work with increased fervor since I was reasonably sure that a major discovery and a major breakthrough in scientific methodology were close at hand. I was convinced that my own strict academic training and experience would eventually find a rationale behind the principles of ancient energy and possibly a key to the Atlantic puzzle itself. According to Plato and the esoteric tradition, the continent had disappeared as a result of the misuse of energy. Edgar Cayce describes a terrible crystal that the Atlanteans possessed through which they channeled destructive vibrations that eventually brought about their own demise. As mentioned earlier, crystals are known conductors of energies in the monuments of Mayan cultures.

Despite Cayce's accounts of the last days of Atlantis, I was unable to understand how an entire population could have destroyed itself through the misuse of energy. Cayce indicates that the Atlanteans knew that the end of their civilization was imminent. He claims that shortly before the final destruction they left Atlantis and travelled into Egypt where they buried records of their civilization. Cayce also indicates that records were hidden in the Yucatan Peninsula of Mexico. Preliminary searches for these records by the Ancient Mediterranean Research Association were made in 1976 and 1977 but neither location yielded any information or the possibility that the readings were accurate.

One of the Cayce accounts deals with the relationship of the

Basques of northern Spain to Atlantis. He indicates that beginning in 29,000 B.C., the Basques fled Atlantis into the Iberian Peninsula when upheaval from earthquakes caused their land to sink slowly into the Atlantic. My explorations in and around Bilbao, in the heart of the Basque country, caused me to suspect that there was a possible connection between the mysterious origins of Cro-Magnon man, Cayce's migratory Atlanteans, and the enigmatic roots of the Basques. Scientists have attempted to demonstrate that Cro-Magnon man came from the East. However, notation-type inscriptions which are mentioned by Alexander Marshack in *The Roots of Civilization* as highly advanced systems of calendrical writing devised by Cro-Magnon and his forbears have been found in the caves of northern Spain. The sophistication of this writing and the similarity of cave drawings all along the northern coasts of Spain and the southern coast of France, led me to believe that Edgar Cayce may have been correct in his description of migrating Atlanteans who moved from the Atlantic Ocean into the north of Spain between 29,000 B.C. and 10,000 B.C. These dates coincided with the arrival of Cro-Magnon man in the area around 20,000 B.C. The origins and language of the Basques are unknown. Although they resemble the Celts to a large degree, they are unique among the Indo-European tribes and may reflect a race whose roots originated apart from all others in the north of Europe. I spoke to many Basque people, particularly a doctor who befriended me, and learned that some Basques ascribed their origins to Atlantis. This idea has been noted in an article in *National Geographic*, August 1968, which stated that there are three theories related to Basque origins indicating that they are: (1) survivors of Atlantis, (2) the last vestiges of Cro-Magnon man and (3) descendents of the mysterious Iberians who once peopled Spain. According to the Basque doctor, Atlantean refugees occupied the caves and protected areas of the Pyrenees Mountains when the island continent began breaking up. They were also escaping floods, since the great layer of ice that covered Europe started a slow melt, causing the oceans to rise more than several hundred feet at the time of the final breakup of the continent as described by Plato in 9600 B.C. My observation at the Caves of Santamamine assured me that such ideas had a basis in fact because the drawings in the caves were not primitive, as in the case of Cro-Magnon man but displayed an advanced aesthetic sense atypical of other prehistoric cultures. Actually, I had never doubted the existence of Atlantis at any time, even in the earliest days, when I hadn't a shred of physical proof to support my hunches. My sensitivity to the ancient energies of Spain provided me with abundant information and an unbridled enthusiasm for my work. Despite the Cayce ideas that Bimini was Atlantis, as well as the fieldwork of divers in the Bahamas, I was reasonably confident that the most important remains of Atlantis were off the coast of Spain. The Gates of Hercules were thousands of miles from the Bahamas and the data overwhelmingly favored the Iberian Peninsula, even though Atlantis may have spanned the Atlantic Ocean in a series of islands, making both locations possible sites. Since the waters in Spain were closed to diving after our search, we were unable to continue investigations to verify that Atlantis lay on the continental shelf near Cádiz. But my correspondence with Paco in the years since the search as well as several visits to Spain indicated that in his diving he observed more cities and artifacts, suggesting that I was on the right track in pinpointing the former existence of advanced cultures on the underwater shelf area near Cádiz.

The Bimini site did not provide me with the clues I sought relative to Atlantis, yet there were similarities to the Spanish site in terms of the size and placement of the blocks. Since I believe Atlantis once existed above water in the Atlantic Ocean, all of the coastal areas adjoining Europe and North America may have once been part of the Atlantis complex. But my discussions with Dr. Valentine during the spring of 1973 in the Bahamas convinced me that the use of ancient energy for underwater archaeology was valid. On an expedition with him to search for underwater remains in the Bahamas, two professional divers tried unsuccessfully to locate the megalithic blocks related to the lost continent while the rest of us waited impatiently aboard ship. Dr. Valentine had observed the Bimini road five years earlier, so he knew the approximate place to drop anchor. The road is a series of megalithic blocks, which weigh many tons, leading underwater

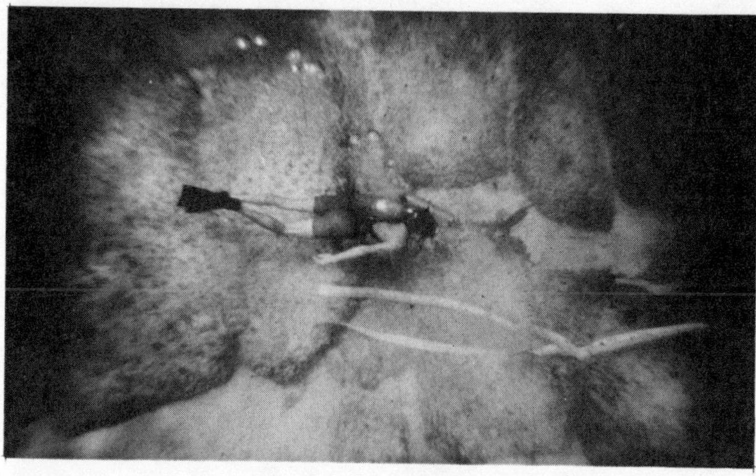

Megalithic blocks found underwater in the Bahamas near the Bermuda Triangle are over 10,000 years old and may be the remains of fabled Atlantis.

from Bimini Island to Andros Island in the Bahamas. Our chief diver, a well-known oceanographer and underwater explorer, could not find the roadway. His photographer was no more successful. Both men had descended at least thirteen times and were physically and emotionally exhausted. Meanwhile, Dr. Valentine and I were concentrating on tuning into the energies in order to pinpoint the location of the blocks for the divers. We finally decided on the most promising spot and pointed it out to the men. At first they scoffed at our suggestions, but their resistance gave way as they were tired and anxious to return to shore. The position for the descent was slightly off the right side of the boat, no more than twelve feet away. On the next dive, they found the blocks. Within thirty minutes they had photographed the site and an hour later we were back on land.

After that experience, I decided to abandon Bahamian research, since I was more comfortable searching for Atlantis in the Atlantic where the evidence in favor of its existence seemed most plausible to me in light of Plato's story and my own research. The Bimini work was continued the following year by

Underwater block from Bimini has a pit and groove construction, an engineering technique not known to the Mayans. Perhaps earlier cultures used energy principles in constructing their sites.

Dr. David Zink, Peter Tompkins, and Count Pino Turolla. The teams used approaches involving both scientific and esoteric techniques. An archaeologist with Dr. Zink, John Steele, who lectured for AMRA (the Ancient Mediterranean Research

Association, which I had founded in 1972) in Spain, reported that such avant-garde methodology was essential in Atlantis research. According to John in an unpublished brief, "As more and more correlations are made between science, myth, and altered states of consciousness, the probability increases that Bimini is one of the most important archaeological sites of the century (for Atlantis research)."

Eventually, I realized that my main quest was not to find Atlantis or to document the importance of archaeological sites. At stake was a new methodology whereby ancient energy would be developed as a tool for the investigation of human origins. The verification of Atlantis in the scientific sense was doubtful without conclusive physical evidence. That evidence depended upon the validation of a new technique to supplement the existing linear view used by most scientists. Atlantis was a vehicle by which traditional scientists might reevaluate their rationale. Consciousness was as important for investigations as the archaeologist's spade. If we can believe the intuitive accounts of Atlantis, then it is possible that the Atlanteans, like other prehistoric cultures and some contemporary primitive tribes, developed a keener perception in order to acquire knowledge naturally available to them through the energies. These possibilities, which appear to have been used in ancient times, encouraged me to extend the scope of my research to include ancient energy methodology as well as the traditional examination of blocks, stones, and artifacts. Atlantis research offered a revolutionary new way to perceive data. I encouraged the use of that perception over and over again as I continued to spearhead an alternative search for our ancient roots.

8. Forewarned Is Forearmed: A Manual for Exploring Ancient Sites

Ancient energy, as I tested it under a series of changing circumstances, was unpredictable. Personal factors and planetary events altered the process of reception, causing inconclusive test results. After a vernal equinox and a full moon appeared during the first week of our trip followed by a lunar eclipse at the end of the second week, the usual vibrations were disrupted at the Mexican sites. A chance meeting with Raul Mendoza at Villahermosa suggested to the group that we were correct in our assessment of the energy upheaval since Raul appeared to understand the relationship between the planets and events on earth. Giant Olmec heads transferred a force to Raul when he placed his fingertips on the stones. These heads were made of granite and were excavated at La Venta, not far from Villahermosa, where we were staying. Many of the heads weighed from 15-25 tons and were carved by a race of people whose origins are unknown. Most of the heads had Negroid characteristics and a few had Oriental faces. Raul had practiced for years placing himself in the force field of the statues where he claimed he actually became part of the same energy that composed the statues. Depending on the cosmos, and his own state of consciousness, he experienced a strong energizing effect from the field surrounding the heads.

Raul helped me to understand that the essence of his body,

and the energy in the heads, differed only in form. By aligning his perceptive capability with the energy mass of the stones, he interpreted the signs and symbols on the faces to be a system of astronomical predictors devised by the Olmecs. Raul supported his theory with evidence he had collected from mathematics, physics, and related sciences. He encouraged the academic community to observe his experiments. Several local professors took photographs of him with the Kirlian process, and Raul claimed the pictures revealed a circle of blue light around his body whenever he touched the heads. Raul was the perfect investigator of ancient energy. His commitment to the statues involved all of his time and he fully understood the principles and practice of using ancient energy to secure new knowledge about the prehistoric world.

On my first visit to the Olmec Park, I was not sensitive to the field of force because of my own highly emotional state, the result of an argument with friends just prior to leaving the hotel. Irritable and depressed, I photographed the statues without regard for their energy-producing potential. I developed the slides in Los Angeles but not one of them came out. Raul explained that the negative alignment of my consciousness resulting from the argument destroyed the circuit between me, the camera, and the statues. He told of a similar experience he had had at Palenque, in the jungles of Chiapas. Several excellent photographers had entered the tomb at the bottom of the Temple of the Inscriptions. For no apparent reason, their cameras failed and they were unable to secure even one picture. Raul said the disfunction was caused by (1) lack of harmony among the men, and (2) the strong energy produced by the megalithic tomb block, a sacred Mayan artifact. They unconsciously produced the interference preventing a clear connection to the force field, just as strong radio waves inhibited all others in the surrounding band. Such problems were familiar to me after our equipment failures inside the pyramid of Cheops, but I had assumed I would have no difficulty photographing the Olmec heads in the brighter light outdoors. Raul said that interference can be even stronger in the park because of distractions coming from above and below the field. For example, he believed that earthquakes emanated from outer space and not from inside the earth. Raul's opinions were those of a layman but I had learned in my investigations to respect all people who manifested experience in certain areas regardless of their credentials. Since Raul indicated that he frequently discussed these issues with the local professors, and since I also noted that he was well respected by the museum director at La Venta Park where the heads were displayed, I felt there was some validity to his theories. The others in our group shared my opinion about his expertise because, in his way, Raul was trying to conduct scientific experiments with the statues in order to demonstrate their connection to planetary activity. He kept careful notebooks, tried to verify his results, and repeated the same tests over a period of months and years.

The concept that heavenly bodies might affect the flow of energy on our planet changed my traditional ideas about the nature of earth movement. It also indicated that human energy levels might be affected by cosmic events. I suspected that astrology, as practiced in antiquity, could be based in fact. The moon influenced the tides, and the sun was vital to human, plant, and animal growth. An article in the *Los Angeles Times* on Sunday, July 16, 1978, mentioned the effect of the full moon on body cycles. Certain illnesses increase during this period and, in general, energy is heightened. Guy Underwood, writing in *The Pattern of the Past*, indicates that dowsers, working at the megalithic sites of the British Isles, pick up positive and negative energy in a spiral as they move along the earth's surface. This energy appears to vary according to the phases of the moon. The ancients also knew that planetary factors could directly affect the human body. Their concern with outer space was evident to me when Raul explained another purpose for the glyphs on the heads. Since they were lined up with the position of the planets, they could be used as predictors to avoid periods of heavy energy when the planets potentially generated destructive energy in the earth plane. Raul explained the correspondence by drawing sketches on the ground in front of the statues to clarify the obscure messages carved on the heads.

Not everyone was prepared, as Raul was, to receive the powerful vibrations. Visitors to the park experienced the full impact of the energies only when certain restrictions were imposed. For example, I felt that eating affected interaction with the force. Interference occurred when the density of food caused the rate of vibration in the body to drop below the higher frequency of the site; the molecular structure of the flow was at odds with the food eaten. The effect was like a boulder dropped into a glass of water. The participant "grounded" him- or herself due to the sudden density. If intake was minimal, he or she stayed in the area of higher energies and easily received the lighter vibrations of the field. Since physical reality was often a product of consciousness and the sensory processes, manifestations at the site depended on energy interacting between the investigator and the environment. Excessive eating often prevented a direct connection to the force field. A light meal eaten at least two hours before exploration was more satisfactory. Fasting was dangerous, since without body density the energy transmitted by one's level of consciousness often rose above the energy of the site, missing the direct connection with the energy field. A clear state of mind, relatively free from thought and emotion, was best. Early morning visits after sufficient rest and meditation resulted in successful experiences. Misuse of energy occurred when the subject was in a stressful physical or emotional state that clouded consciousness and obstructed the sensation of the force field. For maximum benefit, bodies had to be cleansed, balanced, and aligned.

The techniques of tapping energy used by our exploratory team were models drawn from my own work. Investigators placed their hands, palms forward, on the statues, and visualized a flow of white current coursing through their bodies. They were encouraged to walk around the object several times, observing irregularities, and trying to view the artifact in a spatial configuration rather than as a separate entity. I also suggested that they step away from the form and walk back slowly toward it, repeating the procedure until they had a "feeling" for the object and the site. Technical details were ignored during the early stages of research. Following the theory of ancient energy had its own rewards. More than half of the twenty people on our team felt a heightened response, perceived information about the artifact they had not known before, and experienced a surge of personal creativity. La Venta Park is one of several good sites for inexperienced investigators to practice their abilities, since it is accessible, arranged in a pleasant setting, and is part of a high vibrational field.

Any distraction or emotional disturbance potentially cut off receptivity to the vibrations. As soon as possible after contact with a site, I encouraged people to discharge energy in any of the ways mentioned in Chapter 6. Often I suggested that they go back to the hotel and spend time alone in meditation or contemplation. Frequently, I recommended group singing, dancing, long walks, bicycling, and other pursuits where they could involve themselves as participants. Although eating per se was not a discharge mechanism, it did help some people to drop below the frequency of the site and return to a normal level of response. Even sleeping released energy to a lesser degree. My own custom was to light a candle and remain alone in my room listening to music or simply engaging in a brief period of reflection. I usually avoided personal contact during explorations and took at least two days to be by myself before the work began. This enabled me to be a clear channel at the site. Other receivers of ancient energy practiced solitude. Raul lived and worked alone, and Paco isolated himself from friends and family for days at a time. If my energies moved away from what I felt was my peaceful center the return to stability was difficult. During periods of imbalance, I observed that severe emotional problems occurred in some individuals. Misalignment resulted from faulty consciousness or the conflicting vibrations of other people, producing a "see-saw" effect on the body as it tried to regain balance. During this process a series of rapid highs and lows took place; I preferred to avoid this condition by not allowing myself to get into situations which caused it. My mother was aware of this interference with her energy flow and maintained that "anyone who poisoned her environment couldn't come into it."

Frequently, I perceived a change in the force field of my home as soon as I walked through the door. After my experiences at the archaeological sites I was more open to energy centers that can be created at any location where vibrations are intensely discharged. One New Year's Day, after returning from a trip to San Diego, I sensed something was wrong as I entered the living room. There were no physical manifestations to arouse my suspicions, but I learned later that one of my daughter's friends had tried to commit suicide in the apartment the night before. The vibrations were so intense it took three days to discharge them from the environment. During that time my home transferred static from the opposing forces which were deposited in what was normally a clear channel. An energy field can exist in a home or at an ancient site. The intelligent investigator must be aware of these vortexes and his or her body's own reactions to them. Purification and alignment with vibrations is an ongoing process in the investigation of ancient energy regardless of where that energy is centered.

Ireland was a remarkably pure site and an excellent place to begin preliminary studies of ancient energy. Most areas of the island had a higher, lighter vibratory rate, especially at Dun Aenghus, on the Aran Islands, where I sensed a strong positive field of force. Residents of the area acted as if they had little sense of time or space exemplified by old men who stood in the taverns all night long, throwing darts in a slow-paced rhythm for twenty-four hours or more without letup.

One afternoon, while I was investigating the remains of an ancient city on the cliffs at Dun Aenghus, I met several members of the Yeats Society who told me that the great poet and philosopher had been inspired to write his works there because he felt more creative near the cliffs. A similar story was told about composer Manuel da Falla in Spain. It is interesting to note that both men received stimulation from areas near the Atlantic Ocean, the same fields that produced good energy for my own work.

At Dun Aenghus I observed miles of regular blocks and walls near the overhanging cliffs superimposed on a weathered limestone shelf. A major ancient city had existed in the area and I wandered around the remains for days feeling the energy in motion. My step developed a spring like the photos I had seen of men walking on the moon. Irish legends report that in olden

Dun Aenghus Ireland. The Warm Hole is a manmade structure believed to predate man's earliest European origins.

times people could throw a megalithic boulder single-handed from Galway to County Clare, a distance of eight miles. I contemplated the possibility that the buoyancy I sensed might have been experienced in ancient times and influenced the writing or telling of such myths. Perhaps the boulder had become less dense so that it travelled easily over long distances. Like most folk tales, Irish myths have elements of truth in their narratives. One theme that pervades Irish legend is that of big and little people, reflected in such stories as *Gulliver's Travels* and the tales of leprechauns. Whether it was the giant size of the people, or the heightened energies that caused boulders to travel over long distances, we cannot know. Yet hand-carved megaliths dot the Irish countryside and many stones have been taken to museums where they remind visitors that some myths may have a basis in fact.

The myth and magic of Ireland is not all whimsy. Folklore collectors still roam the countryside collecting oral history from the residents who often report seemingly irrational phenomena. In County Limerick, Laurie and I were invited to stay at Castle Matrix, owned by Colonel Sean O'Driscoll. Just before our host left us alone in the castle, we were advised to "be careful of ghosts," since legends relate that the Earl of Desmond was murdered on the site during the Middle Ages. Sean, a leading Irish intellectual, recalled a dinner party where the windows of the library came flying open at the mention of the dead Earl's name. He explained that there were forces in and around the castle that frequently manifested themselves in unexpected ways. Although Laurie and I only remained one night at Castle Matrix in Rathkeale, we sensed the power of the ancient vibrations mentioned by Sean. From the moment we were left alone in the castle by Maur Markham, Sean's housekeeper, Laurie kept hearing footsteps and other kinds of noises, which I did not detect at first. We both knew our imaginations were working overtime after the tales we had heard of ghosts, but there seemed to be an atmosphere in the castle that night which approximated a Shakespearean tragedy based on the intensity of the feeling we both sensed. Laurie and I are both seasoned travelers and have few, if any, fears in a strange location. Yet, by bedtime, we were "frozen" in our chairs when the vibrations started to become very dense. We finally locked ourselves in the bedroom where we slept together on a four poster bed. One or the other of us stayed awake during the night, kept the candle burning, and concentrated in an effort to raise the energies and protect ourselves from an unusually volatile situation, not a normal occurrence in view of the positive forces usually surrounding us in Ireland.

The Irish naturally maintain the force field of the island through their cultural traditions. The land, isolated throughout history, transmits the original aura of early Celtic tribes. Harps, flutes, and drums conduct the energy, along with folk legends, poetry, and dance. The Irish bind past, present, and future into one holistic framework. Like other visitors, I lost my sense of time in Ireland, and became caught up in the eternal idea of a shared ancient inheritance. The "wake," or celebration of everlasting life, reinforces the concept of a joyous transition from the physical to the nonphysical state. Irish preoccupation with genealogy and the study of ancestral origins added to the positive energies I felt on the island, where there was little difficulty communicating with the fields of force.

At the Caves of Newgrange, near Slayne, triple spiral petroglyphs, like those on the Canary Islands, may have expressed the continuity of life embraced by the ancients. It is possible that the early Irish cultures and the Guanche Indian tribe understood the same symbolic concepts. A speleologist, who designed jewelry according to the spirals inside the caves, explained what may have been the function of these ancient petroglyphs. He believed they were a form of communication, designed by the original inhabitants perhaps to conduct the high vibrations of the cavern. For thousands of years sacred rituals took place at Newgrange encouraged, I believe, by the activating principles of the spiral design. In *Earth Magic* Francis Hitching cites a report given by Michael Morris at the Irish Archaeological Research Forum in 1974 which indicates that the spiral form is an integral part of ancient ritual symbolism which is expressed in the etchings on monuments such as Newgrange. The spiral and the concentric circle are the dominant motifs. They symbolize the life energy or rhythm governing the universe of early man.

The ancients may have concentrated on the spiral form and used its symbology in their sacred rites. As a result, the flow of the spiral probably served to accelerate the vibration of their ceremonies and create a power whose force field is still perceived in Ireland today.

As a result of such fields of energy created during religious ritual, my observations at the sites were enhanced since I not only *saw* but *felt* the cultures which had passed through these areas since the dawn of time. Each time I visited the ruins, my clarity and understanding of ancient Irish history increased.

Complex symbols, advanced building techniques, and mysterious writing systems indicated to me that one or more preflood cultures probably occupied the area. I searched for literary evidence to support this idea but was unable to locate any hard data. Each time I returned to Newgrange, I gazed at the wall behind the megalithic boulder that guarded the entrance to the cave. A geologist friend from Trinity College verified that the blocks were volcanic, yet I knew that there had been no volcanos in Ireland. After I learned that the stones were of the igneous type I realized that a great cataclysm must have placed them at Newgrange where the early inhabitants of the site used them to build their cave. Shortly after leaving Slayne, I finally obtained books that supported my ideas, something which frequently happened after a long period of field investigations. These books made me suspect that the volcanic material was the result of a violent undersea explosion in some time past. Since I went directly to the information I needed, moving along to each data source automatically without the use of conscious reasoning, it was my strong feeling that I was again drawn to the right information following a great surge of energy experienced at an ancient site. Research in subjects as vague as Atlantis is difficult, since dozens of academic disciplines apply to the problem and an investigator cannot hope to excel in them all. My ability to quickly find data in fields distant from my own, such as geology, was unusually keen. I believe the energies played no small role in that acuity.

Coincidences such as discovering books supporting my ideas about Newgrange are explained, I believe, by Buckminster Fuller in his book *Intuition*. In the mechanics of nature, everything in the universe is connected with everything else. The energy of Newgrange and the books I discovered were all part of the same energy vortex, which stayed in my consciousness even after I visited the site. The initial contact with the field and the subsequent reinforcement of energy through repeated visits led directly, through some unconscious process, to the data I was seeking. One book, *Geology and Ireland* by Professor Nevill of the University of Cork, shows a geologic map of Ireland in which Atlantis is recorded as scientific fact. Another book, printed in the sixteenth century, mentions a preflood giant of the Bible who gave the island its earliest name, Ogygia. Little by little, I gathered support for my idea that the Newgrange site was built before the worldwide flood which Plato placed at 9600 B.C. and that the volcanic rock probably came from a great catastrophe or volcanic disturbance that pushed rocks up the Boyne River from the Irish Sea into the Newgrange area where the caves were later constructed by cultures who used the igneous material as building blocks. That disaster appeared to be the one referred to by Plato in his Dialogues "Timaeus" and "Critias." The location and style of the caverns were almost identical to other sites in Spain and the Mediterranean, where I perceived strong similarities in the construction of certain monuments. Areas which bordered the Atlantic Ocean, regardless of how far apart they were geographically, may have had a common cultural and geological origin tying them together historically in the minds of some investigators. Scientific support for such observations usually came after I visited a site. My expanded energy potential drew me to the books, artifacts, and other information required for my work.

Many visitors to Ireland share the same peak experiences. The island casts a magic blanket over almost everyone who enters, projecting a vibration that has endured since ancient times, due to Ireland's isolated position in the Atlantic. The field is easy to contact when certain restrictions are observed. Preprogramming about the site must be avoided, a critical factor in dealing with all energy questions, especially in areas that receive heavy public exposure, such as Stonehenge and the pyramids. I suspected that advanced knowledge of the sites placed false energy into the consciousness that blocked actual reception in the field. I noted that pure channels were created by observing, recording, and listening, but not talking or constantly engaging in introspection. I have watched people diligently study guidebooks before their arrival at high energy areas. Later, they failed to understand why they were unable to contact the force. When I explained that they blocked their own field by excessive reading, they gave less attention to the physical process of securing information. Some people forced the flow by continuous conversation or by transmitting conflicting vibrations. It was better to quietly receive energy by adopting a posture of "being" rather than of "doing." Taciturn people, like ancient men, talk little and receive more. As I mentioned earlier, lack of verbal communication is still observed among the Cuña Indians of Panama and other native tribes. When I realized that guided tours blocked receptivity to my own sensory abilities, I made it a practice to lecture on the day preceding investigations. This allowed participants time to examine sites alone without interference from me or fellow workers.

Although Ireland is an ideal spot to perceive ancient energies, Spain, Italy, and other areas of the Mediterranean were also excellent for initial investigations. Researchers were advised to begin work in Europe and later transfer their experiences to the Americas, where sites generally carried a denser vibration. Haiti, like the Yucatan, seemed to have a heavy energy field. Visitors who moved cautiously into the force field were less affected by the voodoo and witchcraft practiced on the island. Haiti's resistant frequencies adversely affected human perception, since they resulted from a native preoccupation with torture and death. While touring the island, I met a young Haitian, Lucien Guierre. He asked me for money to attend business school and suggested that I send him $30.00 a month after I returned to the United States. Since he seemed like a nice young man, I mailed him three payments. Then for financial reasons, I was forced to stop the stipend. Shortly thereafter, he sent me a letter saying he would place a voodoo curse on me if I didn't continue the payment. I was unmoved by his threats, since I refused to believe in the power of black magic on my own consciousness. However, in the months that followed, my luck went from bad to worse. Later, I analyzed the problem and discovered that I probably had not disconnected myself from the force field of Haiti. Inadvertently, I may have been transferring Lucien's energy into my own consciousness even though I was far from him and the site.

The principles which govern ancient energy are not affected by the limitations of time or space. Thus, an energy field may be weakened by the distance of the receiver from the site, or a connection between the vibrations of the center and the subject can continue over thousands of miles depending on certain conditions which reinforce the flow. Artifacts taken from the area, music, or other sensory stimulating factors may conduct the energy, causing it to remain in force if the subject remains open in consciousness and wishes to retain the charge. Such maintenance takes place in a process simulating telepathy, whereby a sender and a receiver can be located in two different geographical locations and still pick up and receive messages. In telepathy experiments, conducted by Dr. Thelma Moss at UCLA, participants relayed widely diverse messages without recourse to conscious reasoning or traditional methods of verbal communication. In these experiments there was a human conductor and a receiver, placed some distance apart, involved in the transmission and/or the reception of a message. In the case of ancient sites, the site itself can be the conductor with only one person, located away from the area, as the receiver.

Energy should be released after the person leaves a site through meditation and other techniques. Artifacts, pictures, and memorabilia from the ruins may cause a resurgence of the field after the participant has left the physical surroundings of the area which may not always have beneficial effects on the researcher after he or she has moved to another location. Conversely, if one wishes to retain the charge it is possible to reinforce energy by keeping in sensory contact with the vibrations of objects that belonged to the site. Returns diminish if reinforcement is not continued on a regular basis. Unconsciously, the frequency is lost or perceptual pathways are closed off. I visited and revisited important areas in order to "recharge" myself with the energy of ancient cultures since the vibrations kept me in contact with certain knowledge needed for my investigations.

When travelling with other people, I selected aware companions. Everything in my environment had to be balanced for successful investigations to occur. Often it was difficult to explain such behavior to friends. I frequently cancelled plans if I sensed a new or disharmonious pattern to their energies. People were "handpicked" for my expeditions, since I feared contamination of my ability to sense the vibrations. Often such travellers were shipped home in the middle of a field experience. On my Guatemala trip in 1975 a Palm Springs socialite left midweek after the group found her behavior incompatible with its purpose. I even refused to talk on the telephone when the energies of the other person were incompatible with my own. It was safer to write letters than risk interference with the clarity of my own channel. American Indians, like members of other tribal societies, uttered a minimum of spoken words. Signals in hand, smoke, and written form often replaced the use of sound as a means of communication. Holy writings preached that the unspoken word affected human events. Speech was a powerful influence on man's behavior. A strong oration could sway the most resistant of audiences. Careful handling of words and intonations had to be observed at ancient sites.

Subtle vibrations, received at frequencies above the wavelength of sound, often damaged bodies and minds. For example, it has been reported that a brain wave can be locked into the alpha or beta state, potentially producing insanity in the victim. Too much energy at the higher vibrations can also affect the endocrine system and the neural pathways, which ultimately leads to a weakening of all bodily systems. To repel such forces, protective techniques have been devised. By closing off psychic channels and remaining centered, some people can form an aura of protection. But more sophisticated practices exist that can be taught through consciousness training similar to that taking place in our own and in other countries. In *Psychic Discoveries Behind the Iron Curtain*, Ostrander and Schroeder report that major world powers have experimented with mind control. Recently, a key reporter for the *Los Angeles Times* was arrested in Moscow for possession of papers dealing with psychic phenomena, which is known to be a potent force in the harassment and control of nations. We can forestall such potential dangers by learning to handle the energies correctly. When we shield ourselves from the denser vibrations, we will be ensuring freedom in both mind and body.

I sensed a need for added protection on my second expedition to Egypt, since I had an overwhelming premonition to cancel the trip. As the leader, I couldn't justify changing the plans of several dozen people, so I went. Unfortunately, my foreknowledge was correct and the obstacles were overwhelming. I was physically, psychologically, and spiritually devastated for three months after my return and my ability to work with the energies disappeared. Prolonged investigations at the sites, added to other sorts of strenuous activities, were a drain rather than a blessing. They depleted the new energy channeling through my body, and I never "caught up" metabolically with the effects of overstimulation. The body's capacity for receiving ancient energy is limited. Overexposure to vibrations at the sites and physical stress could cause me to lose my sense of the natural flow.

Before working abroad I believe the beginning student should conduct some experiments with ancient energy at Indian burial grounds or other local ruins. Successful performance in these earlier experiments will not guarantee carryover behavior in the field, but it may help a new investigator become acclimated to the power of the forces. Many variables at the sites demand great persistence and skill before beneficial effects are realized. At Stonehenge, I was embarrassed when a friend asked me to give him a reaction to the energy field surrounding the megalithic pillars and boulders. I knew that Stonehenge was an area of high energy but I was experiencing perceptual anesthesia. Either the energy was so powerful that I was blocking it out, or the crowded conditions were producing conflicting vibrations and preventing me from picking up the wave. Often I could not connect with a site that other people recommended highly. For that reason I avoided travelling to Israel and the Near East because I did not feel comfortable with the vibrational field as I perceived it away from the site. I had to feel a certain readiness before I moved into new areas. A cosmic clock would be turned on at appropriate times, compelling me to action. Each person has to recognize the signs of his own inner timing and act accordingly.

Harmonious encounters with the force depended on the quality of the experience at the ruins. People eventually came to understand their energy potential as they conducted experiments with growing degrees of success. Revisiting the sites caused sensation and perception to increase. There was a factor of

The author's photograph of a face at Stonehenge England. Can more acute perceptions of existing monuments result in new and unusual archaeological discoveries?

confidence involved in energy experiments which improved each time investigators had success at a site. When people repeatedly experienced the feeling that they could direct the energies, they automatically generated more of the same kinds of experiences.

This confidence resulted in a kind of faith, which I viewed as a focusing mechanism for the vibrations. The stronger the individual's belief structure, the more precise and concentrated his perception became and the greater his ability to direct energies. In a sense, the pyramidal form, like other energy collectors, is no more successful in channeling the forces than the human mind, assuming the mind could be trained to receive the energies and maintain its focus. Working at high energy centers using appropriate safeguards is one means of developing new knowledge, increasing creativity, and providing a stimulating outlet for self expression. The basic safeguards for research into ancient energy are summarized below:

1. Maintain optimum mental and physical health.
2. Practice interaction with vibrations at local ancient sites.
3. Do not preprogram information about the area you plan to visit.
4. Begin work in relatively untravelled regions.
5. Eat lightly before experiments.
6. Transmit less and receive more. (Silence is golden.)
7. Always discharge energies after leaving a site.
8. Manifest a positive aura of protection at all times.
9. Systematically record observations and experiences.
10. Be patient in waiting for results.
11. Travel alone whenever possible.

9. Ancient Energy and the Survival of the Planet

While my study of ancient energy suggests a change in traditional views of the universe, the scientific community as a whole refuses to apply the principles of modern physics to research this energy. Tunnel vision on the part of investigators actually hampers the world community faced with a growing mountain of problems such as the increase in violent crime, air pollution, incurable disease, and learning disabilities. In the past, the university has traditionally taken the lead in helping cure society's ills but by 1970 an impasse had been created, and a chasm grew between the kind of laboratory experiments performed and the effect their findings had when put to the test in the world outside of the university. This was partly due to the nature of the problems themselves which involved a changing consciousness among many people who reacted in unusual or even violent ways to situations involving the environment, law, medicine, government, and human relations. The scientific community's failure to solve some world conflicts may have been based in part on incorrect data. In April 1973, Frank Rhodes, of the University of Michigan, stated in *Time* magazine that the qualities we presently measure have as little relation to the world itself as a telephone number to its subscriber. What I believe the academics overlook is the factor I call ancient energy which, if applied to investigations in all fields, might revolutionize the experimental process.

One of the first areas where I observed the need for radical change was public education. My experience with elementary school students caused me to question my own ideas about what constituted knowledge. Pupils in the classroom frequently gave irrational responses, unrelated to the questions. Often their answers to questions in social studies and the behavioral sciences could not be verified in the traditional sense. My concern about learning strategies led me to investigate techniques of inquiry and critical thinking in which students' intuitive levels of response were as important as their cognitive skills. Formerly "incorrect" answers suddenly became logical when traced to information received via the left, or affective, hemisphere of the brain. Answers weren't wrong. They were simply part of a gestalt or a larger picture that in time, the learner would eventually come to understand. Some students obtained insights from nonmaterial sources, but they were doomed to failure by the nature of the schools and our present methods of evaluation. I felt from my earliest years in education that I wanted to help these students. One of the best ways was to work with fellow educators so that all of us together would remain on a similar track. To this end, I taught dozens of inservice training classes for teachers, developed new materials, and travelled across the United States as an educational consultant in fields related to the inquiry approach in the social studies. This method involved Socratic questioning and an open look at concepts whereby all data was considered valid unless proved otherwise by new evidence. The student was encouraged to seek information, set up hypotheses, and find his own data in response to the problem. This was different from the traditional textbook approach to learning. From the outset, I faced constant frustration, even though such prominent educators as Benjamin Bloom demonstrated that students learned through the intuitive mode, and that such responses could be measured and evaluated. His book *Taxonomy of Educational Objectives: The Affective Domain* was my bible. Yet I was less concerned with measuring the answers than with creating a classroom environment that would encourage such information to flourish.

Just as the ancients used vibration to trigger energies, I employed color, light, sound, and other sensory devices as components in my teaching strategies. For example, with one particularly difficult class in world geography, I used different sensory mechanisms each day. When the students were reading about France, "Gaite Parisienne" was playing in the background on the first day, while the class ate brioches and French toast on the following morning. As we moved through the unit of study we often had simulation games of a night in the Latin Quarter, or painted surrealistic scenes like the French Impressionists. In short, I saturated them with the vibration of France and they more easily learned the subject. Students were encouraged to explore questions on their own time in the way that was most effective for them. There were no right or wrong answers but simply hypotheses and explanations. The process rather than the content was changed. Since my classes did as well or better on their examinations than other students, I knew that the energy approach to learning was valid. This was particularly true in the study of history since I frequently recreated the feeling of an ancient site by bringing in artifacts of the area, showing slides, and developing an atmosphere as close to the real experience abroad as was possible, within the confines of the classroom. During these units of study, I worked to some degree in training students to sense the energy of the ancients, whereby they put themselves into the same mind set as the prehistoric people they were studying.

I also made a study of ancient education. Teaching methods before Greek and Roman times bore no resemblance to present techniques. Bards and minstrels sang lessons to students in outdoor environments. Classes were small and only the male elite was allowed to take part. Pupils interpreted the information, largely folk legend and oral history, according to their natural instincts. Frequently, they changed the instructor's stories by adding new characters or recreating other situations and problems for the hero in the story to solve. Sometimes they devised similar tales, centered around accepted parables or principles, like the moral lessons found in Aesop's Fables. There was a natural tone to the kind of learning which took place in antiquity based on certain universal or ethical principles. Up to the time of Aristotle, there was little formal education except for a rudimentary kind of curriculum which included theology, mathematics, and mythology. The material and the nonmaterial world was viewed as part of the same reality. I used these ancient approaches to learning in my classroom along with traditional methods. These proved effective, especially with students who previously were discipline problems. A fragmented class became harmonious when I taught a lesson outdoors or played music to raise the vibratory rate. When problems became insurmountable, I used diverse motivational techniques to change the force field immediately. The results were overwhelmingly in favor of an energy approach. I even sensed what my pupils were thinking and tried to anticipate their responses. One of my favorite statements was, "I think I hear you saying . . . ," implying that I wanted to make sure that my perception of their answers was correct. By appealing to the affective part of students' consciousness rather than the purely cognitive domain I allowed for a wider range of responses than is usually found in the classroom.

The holistic approach to learning and the use of ancient energy comprised my solution to the crisis in the classroom. I worked at the method for ten years but eventually was discouraged at the level of teacher training, the heart of the change-making process in education. Most adults had closed off their perceptual processes to all data except the observable. The teachers were willing to try the processes I suggested, especially after I provided them with prepared lessons, materials, and resources. However, they could not succeed in practicing the approach because they did not understand that they had to become a part of the process themselves in order to make it work. They read the lessons, taught them, and immersed themselves in new concepts such as J. P. Guilford's structure of intellect identified in his book, *The Nature of Human Intelligence*. Guilford believes that there are many factors of intelligence rather than one general intelligence. Although he makes a strong case for changing educational methods, teachers remained uncomfortable with such new strategies. Students sensed that the teachers were ill at ease with the approach. Before long, the easier didactic method of teaching was resumed.

Ultimately, techniques of ancient energy as teaching methods were abandoned as a result of three major problems. There was no sure way to verify intuitive data using present instruments of evaluation. Second, teachers resisted practicing new methods, totally different from established techniques. Finally, administrators failed to approve lessons that did not correlate with mandated curriculum at state and federal levels. The process I encouraged on a nationwide basis was doomed to failure before it reached students in the classroom.

A few enlightened educators sensed that drastic changes were needed in the schools. Most, however, abandoned the energy approach, creating dozens of new schemes with impressive titles such as individualized instruction, open classrooms, and flexible scheduling to remedy learning problems. These methods, heavily laden with the products of technology, were like frosting on a rapidly decaying cake. They failed to bring about improvement in reading scores and basic skills. There was a mountain of new information to be learned, and teaching strategies failed to change along with what was really happening in the world. Since science did not always keep pace with society's changing concepts, a discrepancy occurred between established doctrine and the nature of everyday reality. In the *The Structure of Scientific Revolutions: An Epitome,* T. S. Kuhn points out that science collapsed after major conceptual shifts occurred and had to be rebuilt on other structures. It was pointless to revamp old methods. Something totally new had to be developed in both process and content before education could be improved. I believe the answer is to develop pupil consciousness to the point where the students receive information from the energies and apply that information to the vast body of technological data already available. All that is needed to field test the approach in the schools is a crash course in teacher education. This retraining never took place. Instead, blame was placed on the overcrowding in the schools, problems connected with administration, and lack of funding. My idea, unsupported, was discarded in its inception.

In order to correct the current situation in our schools, I recommended that change begin in the selection of school sites and in the plant design where choices could be made on the basis of aesthetic considerations as well as utilitarian ones. The school environment must encourage and heighten the energies of students, if we are to achieve a greater potential in learning. Attention should be given to form, color, light, sound, and other agents which stimulate and maintain a higher vibratory flow. Teacher competency must also be improved in the areas of sensitivity and awareness development so that they may communicate more effectively with students whose consciousness, in general, may be more highly developed than the teachers themselves. A total change is needed in the structure of the curriculum at all levels of instruction to provide for new subjects and new teaching approaches which take into account such areas as the development of energy potential, the use of intuition, the fostering of creativity, and the importance of expanding conscious awareness. These suggested approaches would cost little and achieve much, but since their development would signify a radical departure from established methods, such changes are not likely to occur, at least not in the foreseeable future. I believe that an inverse ratio has characterized the schools: the more technological improvement introduced into the school environment, the less learning that has taken place. Since solutions to problems have not been found in traditional approaches to education, and since educators have failed to develop other innovative strategies, the status quo has continued.

A few improvements were made in the preschool where such educators as Maria Montessori saw the wisdom of teaching both rational and non-rational concepts, an idea also suggested by Professor Morrison at MIT and Daniel Bell who encouraged society to "think wild," as mentioned in *Human Behavior* magazine, November 1973. Creativity itself was inspired by that moment of insight when the inventor said "aha." Such random occurrences could be encouraged, regulated, and measured in students from an early age by developing student awareness to concepts and ideas different from the traditional model educators have established. I firmly believe that process must take precedent over content, since content itself is a changing flow of energy that varies from person to person. Many students resisted facts taught in schools when the information relayed was incompatible with the stream of energy flowing through their own consciousness. Thus they were doomed to failure at the outset, within the confines of the system.

Dr. Jean Houston, a pioneer in the field of humanistic psychology, demonstrated that dozens of untapped potentials exist within each individual. These potentials are blocked off by the vast amount of unimportant information disseminated to students, most of which does not remotely interest them. According to Dr. Fritjof Capra, the ability to learn is determined by the level of consciousness of the observer. John Goodlad at UCLA pioneered educational reform in such books as *Looking Behind the Classroom Door*. He proposed that educational institutions in the year 2000 provide computerized neighborhood learning centers where students could come to learn what they were interested in knowing, and nothing more. In other words, no formal training would be given in the basic subjects like reading and mathematics, but instead students would choose their own areas of study according to specific interest. Since pupils already knew certain concepts and information, they wouldn't relearn the data in a formal school situation. Education in the future might even require that students travel to areas where they sensed attunement. Visits to high-energy sites could greatly expand their potential for new knowledge.

I believe education must shift emphasis from a technological curriculum to a process-centered curriculum whereby the individual is the center of the learning experience rather than the spectator. Technology would not be abandoned. On the contrary, a richer and fuller understanding of how to live within the mechanized society would be possible. Such revisions in our educational process have been suggested by Alvin Toffler in *Future Shock* and Charles Silberman in *Crisis in the Classroom*. Many books emerge with such theoretical ideas. Yet rarely do these notions filter down to the classroom situation because there is no practical way to apply such ideas.

Change in the educational system is not developed by the

schools themselves. Parents and groups outside the institution initiate reform. If the community were trained to recognize the principles of ancient energy, it might assist young people in their ability to learn, and this community support would eventually be transmitted to the schools. Such retraining programs could take place concurrently with the education of students. Family and friends might support radical approaches to learning if they were given the opportunity to experience the success of such methods first hand. Awareness among members of the community is most critical in dealing with unfamiliar subjects like ancient energy. Community education could focus on areas dealing with meditation techniques, parapsychology and psychology, the philosophy and ideas of Eastern and Western civilization, and subjects related to the function of the mind as it pertains to our inner and outer environment. A complete training program might require field experiences in high energy locations, visits by the instructor to assess the "feeling" in the home, and interaction with religious, educational, and civic leaders who support a retraining program based on the expansion of consciousness. Eventually large numbers of students might insist on a new kind of education, forcing grassroots reform that could reverse the disastrous path on which education has embarked. In time, a new system of learning would replace the old, based on raising perception and evaluating knowledge from energy sources used by the ancients. Its effect would totally restructure our educational institutions in all basic subjects and levels of instruction. Suggestions for ways parents can encourage "other ways of knowing" in young people can be found in Appendix C.

Environmental sciences, like education, have been in crisis. Concepts of ancient energy applied to problems of air pollution and overpopulation might avert planetary disaster. For example, in *The Closing Circle* Barry Commoner, suggests that our actions as world citizens are always interrelated. When we pollute the environment in one part of the world, it immediately affects the other part, since balanced ecological systems are undermined by their weakest link. What seems most critical to me is the need to raise the awareness of all world citizens to the point where we are conscious of our collective responsibility for maintaining harmony and balance in nature. In other words, the issue is not the physical destruction to the environment but rather the development of a new sense of morality among people before such destruction occurs. Before the decline of the Roman Empire ancient people were blessed with less technology and smaller populations. Nonetheless, some of the natural concepts they embraced about man and the land might be helpful to review today, such as an understanding of the rhythm and order of the universe and the effect of form, light, color, and space, as they affects energy flow. Vibrations are maintained through the ages in what I call a collective world consciousness, but the most decisive change may have occurred near the end of the Roman era, 396 B.C., when vibrations became heavy and dense with the crumbling of traditional values and the destruction of the physical world as it existed throughout antiquity. When Rome fell, it was necessary for people to concentrate on purely physical survival, ignoring concepts known by the early people relating to the control and use of energy. This necessary preoccupation with the physical environment may have caused the vibratory rate of the energies on earth to become heavy and dense and at odds with the higher, lighter natural flow of the cosmos. It is my opinion that man and nature must return again to a higher energy level by realignment of the natural forces. Only then can technology benefit global purification.

We are all responsible for ecological balance. Waste materials dropped into the ocean near Spain arrived in the Mexican Gulf Stream days later. Byproducts of our supertechnology have residual effects far beyond the obvious smoke and pollution. Experiments with nuclear power, spray pesticides, and cloud seeding have affected energy balance in every part of the world. The depletion of natural resources has resulted from our unwillingness to examine higher energy sources, based on the knowledge of the ancients and the principles of physics. As Pierre Teilhard de Chardin says in *Activation of Energy*, for the first time in the history of the earth, man must organize the maintenance, distribution, and progress of the energy around him, without which he cannot continue to survive on this planet. Thomas Kuhn points out (*Time*, April 23, 1973) that scientists in all fields must be reeducated to perceive the energy factor that coexists with our physical reality. Our society's constant preoccupation with the depletion of physical resources reinforces the incorrect idea that abundance is created by increasing the amount of physical energy available or rationing what is currently available. There are other ways to stimulate conservation, based on the principles of energy and space. Buckminster Fuller, like the ancients, created a holistic structure called a geodesic dome to conserve human resources. The dome was built of hundreds of pyramid shaped panels which Fuller believes help to concentrate the energy flow. People of antiquity also built structures to encourage natural vibrations and channel them into living microcosms of the force. Pyramids, spirals, tetrahedrons, and other geometric forms were the building blocks of prehistory. Domes, beehive shapes, and dolmen architecture also had functional uses to channel the energy of the universe. The ancients had an awareness about ecology that was based on natural law rather than the principles of supply and demand.

Our society has been adversely affected by the inability of construction engineers to use the principles of energy in planning and developing housing of all types. In fact, most cities in the United States would be considered alien in appearance to the artistic tastes of man. In Europe, the oldest buildings were left standing for beauty and remembrance. In our own country, the practical has been substituted for the aesthetic, and this tendency has inhibited our energy potential and dehumanized the environment. Beauty, in all areas of our lives, helps to stimulate the vibrational flow. Ancient Egyptians used henna and other cosmetics to enhance the human form. Our present concern with the "body beautiful" simulates the physical perfection worshipped in the ancient world. What we called physical form is really only moving energy occupying space and is dependent on the perception of the viewer for its reality. This was pointed out by Rudolph Arnheim in comparing works of art. The expression "Beauty lies in the eyes of the beholder," is actually true. A person who sees beauty in someone not normally considered beautiful actually perceives that person with faculties beyond the normal five senses. As we raise our consciousness and come closer to perceiving natural beauty, our evaluation of beauty becomes more accurate and the flow of energy increases. There is a beauty that also comes from living simply close to things of the earth. I believe the adopting of a freer, more natural lifestyle demonstrates an inner desire to come into harmony with the flow of energy in the universe.

Radical changes in medicine have occurred as a result of the application of ancient energy to disease prevention and control. At the Center for the Healing Arts in Los Angeles, doctors have worked with consciousness as it affects the flow of energy in the human body. Other practitioners diagnose illness by "sensing" what is wrong rather than adhering rigidly to the results of physical examinations and laboratory tests. Some practitioners have influenced present medical systems by suggesting that acupuncture and other means of balancing body energies could replace traditional techniques of healing. Particles of energy that should be flowing in balance at all times travel in and out of our bodies. The operation of this essence is governed by noncausal laws. When it becomes misaligned, it can sometimes be corrected with nontraditional approaches, such as biofeedback, which teaches control of the body's autonomic response by redirecting energy coming from various levels of consciousness.

Contemporary medical practice is based on physical techniques of diagnosis and treatment. Traditionally, the body has been viewed as an efficient machine rather than a moving mass of energy that varies among patients and in the same patient tested under differing circumstances. For example, hypoglycemia, or low blood sugar, does not always register on glucose tolerance tests since it is sometimes caused by violent shifts in hormonal balance in the endocrine system, one of the body's physical receptors for the energy mass. Sensitive people experience violent swings in their metabolism as they unconsciously draw denser energies into the body. If doctors were to realize that some illnesses result from an imbalance in vibrations, the accuracy of diagnoses might increase markedly. Retraining an entire medical community of the future to recognize energy flow as a factor in causing illness would mean bringing the human race back into natural alignment.

My understanding of the effect of vibrations on the body radically changed traditional views of diet and health. Slowly I realized that red meats and fats, in excess, depleted my flow of natural energy and increased the possibility that I might suffer heart attack and other illnesses. During research, I rarely ate "balanced" meals nor did I take medications, which had a reverse effect on my own system. Also, drugs and other alien materials introduced into the system upset natural energy flow and produce symptoms. Rather than preventing them, antihistamines caused me to develop allergies. Unnatural substances entering my body upset the flow of energy, reversing normal polarity. Such imbalances, uncorrected, weakened glands, nerves, and muscles not only in my own body but in large numbers of people I observed. The ancients understood the energy quotient of the body and its relationship to food intake. Simple foods such as fruits, nuts, and cheeses were the main staples. I suspect that throughout the history of civilization, there may have been a correlation between cultural decline and the type of diet eaten. I base my hunches on the observation that highly developed societies, such as the Egyptians and Greeks, had a light, relatively meat-free diet. In Roman times, meat was the main staple, especially in the later empire at the time of the barbarian invasions. During the Middle Ages, monastic cults returned to the lighter diet that helped maintain personal and spiritual alignment with the higher forces. Since the body is a receptor of energy, it responds to certain foods whose physical and chemical properties encourage the flow. Although natural resources and custom play a role in the dietary habits of cultures, I believe there is more than a coincidental factor involved in the eating patterns of the earlier Mediterranean societies and the decline and fall of their civilization. For a list of foods recommended to encourage energy flow, see Appendix B.

Other basic institutions incorporated an understanding of the principles of energy in the ancient world. Law was interpreted according to the natural rules of justice. Punishment was a function of religious bodies, or the family group, which tried to appeal to the innermost morality of the guilty. Less complicated ancient cultures, such as the Maltese and the Cretans, probably had little need of jurisprudence since they remained centered at a common level of awareness, due to the natural conditions under which they lived and the simplistic nature of their society. Since death was not feared but seen as part of the same energy cycle, capital and physical punishment for serious crimes was acknowledged as just. The code of the ancients was obedience to the law of natural harmony, or the pure essence of the universe itself, a concept which we might revive by raising and harmonizing world consciousness. If present judiciary policy coincided with such universal principles as love, joy, and harmony, justice might be more equitable for the vast majority of sensitive people who often turn to natural laws when man's laws appear to be unjust. Since such a defense is not acceptable in the courtroom, conflict occurs between people who operate from higher levels of consciousness and those who are bound to the mass consciousness of the existing judicial code. Physicist Herman Kahn states that the evolution of society potentially produces the devolution of the individual because it is inconsistent with the natural order and harmony of the universe. When people in higher consciousness act, they are obeying an intelligence which to them is the law. The reconciliation of man's laws with universal laws could occur through the development of a shared spiritual consciousness, but only if it were acknowledged by our judicial system.

Finally, energy generated in the courtroom could influence the type of decision reached. The interaction of vibrations between the judge, jurors, and litigants could supersede the importance of physical evidence in decision making. I believe the issues dealt with in our courtrooms are not a matter of guilt or nonguilt, but result from interpretations of the law administered by people whose decision-making consciousness is limited to physical evidence and a printed code, which often lacks sensitivity to natural needs and concerns of humanity at large. The application of ancient energy to law could greatly increase opportunities for an equitable system of justice in the future.

Criminal codes must also be reexamined, since violence is known to occur when the body fails to release a rapid influx of dense physical vibrations. The frequency of murder and crime increases during the time of the full moon, when cosmic energy is at its height. I believe most methods of criminal reform have failed because such ideas have not been considered. If society were retrained to properly handle energy by learning to discharge negative vibrations the crime rate might be lessened. Criminals could be people who have run out of ways to release their emotions. Thus the efficient operation of society depends on a harmonious flow of the energies and a balanced mass consciousness. Through a system of global education, such possibilities could be realized, even though it might take generations before the majority of the world community understood the basic principles involved.

Other aspects of the life experience could be enhanced through an understanding of the action of force fields on our basic lifestyles. By concentrating awareness training within the family group, we could disseminate information about energy more widely as individuals entered the community at large. Energy management should be a part of the educational process from infancy on. Children have always had a natural ability to understand universal alignment. As they respond to our technological environment, however, many of their centers of consciousness are inhibited, ultimately resulting in a complete blockage of all but the most basic physical channels. A family group can learn to transfer energies at higher levels of awareness creating a vibrational mass greater than the force of any one of the individual members. As we search for alternate energy sources in traditional ways, we have not considered human potential as the most natural option. But by teaching people the practices of the ancients, we could guarantee a more successful continuity of civilization. Without the practice of such ideas, we will end up in total energy chaos. As physical reality crumbles our institutions will not respond to violent energy shifts. Whenever the the universe experiences changes in its vibratory rate, violence occurs simulating the mythical Phoenix bird who rises out of the ashes every five hundred years to recreate a new reality. Examples from the past and hopes for the future do not guarantee completion. Our guideposts must be the here and now of cosmic balance and personal energy control, governing agents for our own lives.

Amid the chaos, there are people who attempt to harness the energies and tamper irresponsibly with the human mind. When consciousness is manipulated, cultures fall to lower, denser levels of awareness. External control of human perception can cause confusion and spread misinformation among individuals and groups. Entire societies have been conquered through mind control, or led to commit acts contrary to their higher natures. Adolph Hitler, according to biographer Trevor Rayvenscroft in *The Spear of Destiny*, was involved in occult practices which he used to justify mass murder and atrocities perpetrated on the Jews and other minority groups. In fact, Hitler practiced mind control and the manipulation of consciousness to such a degree that many of his victims went willingly to their slaughter.

The potential dangerous uses to which energy has been put have resulted in continuous suppression of such information by a small minority of political and religious leaders who believed they were "keeping it safe" for society at large. Such secrecy and misinformation about ancient energy has hampered society for thousands of years.

While man throughout the ages has worked to develop his physical and spiritual concerns, he may have consciously blocked out his relationship to the forces. This may have been due partly to religious sanctions against such matters which were written into the Bible and permeated the literature of history in such stories as Joan of Arc or Santa Teresa de Avila. The term "witchcraft" was applied to any phenomenon that differed from the traditional experience of society. These fears were expanded over the centuries until the subject of energy became unmentionable and unthinkable, the work of the devil. Today we are returning to these concepts in a more positive manner for the first time since they flourished in the ancient world. People are even considering the possibility that we may be able to understand such ideas as immortality. An exhibit reinforcing such ideas called "Continuum," was mounted at the California Museum of Science and Industry. The show also emphasized that energy is indestructible and that perception is not limited to the five senses. Other channels of public information are beginning to awaken the sleeping consciousness of millions of people to the true reality of the life force.

Although people should be educated in the principles of energy from birth onward, the process may be hampered by fear, lack of information, and apathy about manifestations that are felt but not visible. A system of global education is needed so that people of all ages may understand the concepts of ancient energy. Just as education on birth control has to be global, an entire world community must be reeducated to learn the lost knowledge of antiquity, suppressed for the past five thousand years.

Education of government officials to the realities of dealing with such a potent force would have to be conducted since it is easy to misuse energy by the conscious or unconscious transmission of a negative field. The politics of energy research may be the most difficult to achieve because nations could learn to direct the vibrations in ways which might be more powerful than the hydrogen bomb in their effect on the human mind. Only the most carefully trained people should be involved in such projects who are also oriented toward spiritual concerns and able to make decisions based on information channeled at the higher levels of awareness. Naturally it is difficult to conceive of a selection process for such people placed in dangerous leadership roles since the criteria for their positions would be based on attributes of personality, perception, and personal belief rather than on college degrees or skills. Even if standards were developed by concerned citizens for those in control of energy research, constant monitoring would be necessary to ensure that they did not lose the original criteria required for their jobs since spiritual and psychological traits rarely remain constant. The use of ancient energy, accompanied by strong spirituality, must serve selfless purposes. Otherwise, the knowledge could become a dangerous weapon in the hands of people who would manipulate and misuse it. Enlightened leaders of the communications industry must be made aware that tremendous surges of energy are foisted on the public through the media. This was demonstrated in the summer of 1978, when the screening of one television program affected the entire energy balance of the country, reducing millions of viewers to the vibratory level of the show. The four part series *Holocaust* depressed the population wherever it was aired. It would have been possible to reenact the barbarism of Hitler's Germany without screening four consecutive nights of mass murder and brutality. Other ways to raise mass consciousness to prevent such disasters in the future could be accomplished by using a more positive, if not a remedial approach, to the problem. When negative concepts are transmitted, they bring about a recurrence of the actions we wish to prevent. In energy, like attracts like; what we transmit is what we receive. This radical new idea might change the entire focus of the media, influencing it to disseminate more positive vibrations to the public. Thus, good news would be reported with the same intensity as the constant stream of tragedy that blankets the front pages of our newspapers and magazines.

Many frauds and amateurs have entered into the study of ancient energy, giving the whole field a mystical connotation. Yet some prominent scientists and philosophers have suggested that we retrain society based on the new paradigm of energy. Abraham Maslow said that we must measure mystery, illogical thought, and contradiction along with transcendental experience to get a complete understanding of the universe. Such measurement, taught from birth onward, may help us to realize the quest of Emerson, who said that "generalization is an influx of divinity into the mind." That divinity, according to Maria Montessori, is the environment in which the individual is spiritually and materially immersed. Society must build an appropriate model so that citizens are able to understand automatically the concepts of cosmic forces because they are unconsciously bombarded by such stimuli at all times.

Ultimately, our planet may not survive unless we realize that our past viewpoints have been illusions, alien to the basic nature of the universe. We have lived in a part of the whole. By returning to the energy principles of the ancients, we will raise the curtain of reality. Eventually, we may surpass their knowledge and even get a glimpse of the world beyond. The life force is our basic thrust toward eternity, propelled by the essence of ancient energy. No longer can we choose the direction we will take. Humanity is on an irreversible course leading toward a new cosmic age. The birth pains of revolution are difficult, but the product and potential of the new society will far exceed the accomplishments of the past. We are venturing toward the wisdom of an age-old force which will bring us from the darkness to the dawn. As Cat Stevens has written in a song, "Morning has broken . . ."

Glossary of Terms

AFFECTIVE—occurring at or affecting feelings and emotions.

BIOFEEDBACK—an instrumental technique for self-monitoring of normally unconscious, involuntary body processes such as brainwaves, heartbeat rate, and muscle tension that can result in a degree of conscious, voluntary control of the process.

CLAIRVOYANCE—extrasensory perception of physical objects or events as distinguished from telepathy which involves the ESP of thoughts.

DEMATERIALIZATION—rapid disappearance of a physical object possible due to its loss of specific gravity.

DOLMEN—a monument consisting of several megaliths arranged so as to form a chamber; usually regarded as a tomb.

HOLISTIC—pertaining to the whole, encompassing a complete view of what appears to be a single isolated situation.

INTUITION—the ability of knowing, or the knowledge obtained, without conscious recourse to inference or reasoning.

KIRLIAN PHOTOGRAPHY—a type of high-voltage photography used to measure the energy field surrounding people, objects, or plants.

LEVITATION—the purported raising of objects or bodies in the air without apparent physical means.

MANTRA—a sound frequently used to induce the meditational state.

MEGALITHIC—pertaining to extra large stone objects which are natural or manmade.

ORACLE—the medium by which a god reveals hidden knowledge or makes known the divine purpose; also a place where the revelation is given.

PETROGLYPH—a prehistoric carving on a rock.
PHALLIC—pertaining to male or female sexual organs; generally used, however, in reference to the penis.
POLARITY—the quality or condition inherent in a body which exhibits opposite, or contrasted, properties or powers, in opposite, or contrasted parts or directions; the having of poles.
POST AND LINTEL—an ancient method of construction using posts placed into the ground with lintel stones laid across the top of the stones, which are of massive proportion.
PRECOGNITION—prediction, or knowledge of future events that cannot be inferred from present knowledge.
PSYCHIC—describing paranormal events and abilities that cannot be explained through established physical principles; securing information at an intuitive level of response.
QUANTUM—an elemental unit of energy which reacts according to the principles of quantum mechanics.
SYNCHRONOUS—two or more events which are concurrent in time.
TELEPATHY—extrasensory perception of another person's mental state or thoughts.

Appendix A. A Guide to Archaeological Sites for Beginning Investigators

The following table lists archaeological sites with high-energy potential, subjectively identified by the author on the basis of individual and group research over the past ten years. Beginning investigators should study the list of suggested practices at the end of Chapter 8, and consult the list of sites about details relating to location and special requirements of the area. The phenomenon of ancient energy is not consistent and is dependent on the site environment, the consciousness of the researcher, and his or her belief system, as demonstrated by examples given throughout the text. This listing is partial, since some powerful force fields are inaccessible or have not presently been identified as major sites. Megalithic monuments are spread throughout the area, but Stonehenge is the most important structure. Thus, for each country listed only the most significant areas have been identified. Europe, the Mediterranean, the Americas, and the southwestern United States are all important vibratory centers where experimentation can be conducted at lesser known sites, depending on the investigator's connection to the fields of force.

AREA: EUROPE

COUNTRY	SITE	RATE of VIBRATION	TIME NEEDED to EXPLORE	NEAREST MAJOR CITY AND AVAILABLE TRANSPORTATION to SITE	SPECIAL FEATURES	ADDITIONAL REQUIREMENTS
SPAIN	Caves of Nerja	High-polarized	2 days	Malaga; bus, car	Inscriptions, artifacts, Natural formations	Flashlight, good walking shoes
	Caves of Santamamine	Average-negative	1 day	Bilbao; car	Inscriptions, artifacts, wall paintings, petroglyphs	Flashlight; fruit juice or other quick energy foods
	La Palma, Canary Islands	Average-positive	2 days	Tenerife; plane, boat, bus, car	Lava flow petroglyphs, Guanche Caves	None
	Cadiz	High-positive	4 days	Seville; bus, car, train	Underwater artifacts, Roman roads, caves and ruins, roads	Scuba equipment
	Granada	High-polarized	3 days	Malaga; plane, car, bus, train	Alhambra, Roman and Greek ruins, evidence of Islamic culture	None
IRELAND	Aran Islands	High-polarized	3 days	Galway; boat, donkey cart, bicycle, car	Cliffs at Dun Aenghus, ringed forts, prehistoric remains	Good walking shoes
	Caves of Newgrange	High-positive	2 days	Slayne; bus, car, foot	Megalithic boulders, petroglyphs, Ancient writings	Flashlight
FRANCE	Caves of Lescaux	Average-polarized	2 days	Biarritz; car, bus	Wall paintings, petroglyphs	None
	Lourdes	High-positive	2 days	Toulouse; car, bus, train	Cathedral, sacred rituals of immersion	None

	Cerveteri	High-polarized	2 days	Rome; auto, bus	Etruscan tombs, museum	Flashlight
ITALY	Pompeii	Average-polarized	3 days	Naples; auto, bus	Tombs, wall paintings, homes of the ancients, museum	None
	Syracuse	Average-negative	2 days	Catania; car, bus	Cave dwellings, Roman ruins	High-energy food, walking shoes
ENGLAND	Stonehenge	Average-negative	1 day	London; bus, auto	Megalithic ruins astronomical alignments	Umbrella or other rain gear
	Hagar Qim	High-polarized	1 day	Valletta; bus, auto	Megalithic ruins, phallic symbols altars, inscriptions	None
MALTA	Hypogeum	High-positive	1 day	Valletta; car, taxi	Sacred oracle, megalithic ruins, artifacts, inscriptions	Whistle, pitchpipe, or other vibrational instrument, tape recorder
	Dar Galam Cave	Average-negative	1 day	Valletta; bus, auto	Animal remains, human remains, artifacts	Flashlight
AFRICA/EGYPT						
EGYPT	Giza	High-polarized	2 days	Cairo; taxi, auto	Sphinx, valley temple, Pyramids of Cheops, Chefren, Mycerinos	Whistle, or pitchpipe, other vibrational instruments
	Luxor	High-negative	3 days	Cairo; train, plane, bus, auto	Valley of Kings and Queens, Ramses Temple, Karnak	Good walking shoes

COUNTRY	AREA SITE	RATE of VIBRATION	TIME NEEDED to EXPLORE	MEXICO NEAREST MAJOR CITY AND AVAILABLE TRANSPORTATION to SITE	SPECIAL FEATURES	ADDITIONAL REQUIREMENTS
MEXICO/YUCATAN	Chichen Itza	High-negative	2 days	Merida; bus, auto	Pyramids, sacrificial well, observatory, Mayan ruins	Good walking shoes
	Uxmal	High-polarized	2 days	Merida; bus, auto	Pyramids, nunnery, Mayan, ruins, inscriptions	None
	Tulum	High-polarized	1 day	Uxmal; bus, auto	Mayan temples and ruins	Good walking shoes scuba equipment
	Isla Mujeres	High-negative	2 days	Cancun; ferry, bus, auto	Female shrine, Mayan ruins	Scuba equipment
	Palenque	High-polarized	2 days	Villahermosa; auto, bus	Temple of the inscriptions, Pyramids, Mayan ruins	Flashlight
	Villahermosa	High-positive	2 days	bus, train, plane	La Venta Park, museum	None
CENTRAL AND SOUTH AMERICA						
GUATEMALA	Chichicastenango	High-polarized	2 days	Guatemala City; bus, auto	Living Mayan descendants, sacred ceremonies	Tape recorder
	La Democracia	High-polarized	1 day	Esquintla; bus, car	Giant heads, museum	None

PERU	Machu Picchu	High-negative	2 days	Lima; bus, auto	Inca riuns	Good walking shoes
ECUADOR	Ingapirca	High-positive	1 day	Cuenca; Auto, horsebeck	Inca Fortress	Good walking shoes

Appendix B. Foods That Increase Energy Flow

The following diet is part of a total program that may increase energy flow. Vegetables, meats, starches, and other selections on the approved list should be eaten in three or more balanced meals daily. Foods on the marginal list can be added from time to time. The diet is high in vitamins and low in fats to keep the endocrine system and liver cleansed of fatty materials that could alter the flow of body hormones which control the actions of nerves and muscles. It is recommended that you consult your family physician before limiting yourself to this list.

The emphasis in this high-energy diet is on natural, fresh foods, low in saturated fat. All fruits and vegetables should be eaten raw whenever possible. Almonds are a perfect snack food, since they contain most basic vitamins and minerals. Three to four cups of coffee or tea, taken daily, stimulate the adrenal glands and may help increase energy flow. Small meals eaten every two hours prevent violent swings in body metabolism, but hamburger stands and snack bars where the menu is limited to fatty foods should be avoided. Abstain from taking medications unless absolutely necessary, and substitute hot baths, mineral rubs, massage, and other forms of body treatment that raise the vibratory rate. The use of alcohol and cigarettes is discouraged. Daily meditation and other forms of consciousness-raising techniques should accompany the lighter food intake.

APPROVED	*NOT APPROVED*	*MARGINAL*
Low-fat or dry-roasted nuts	White breads	Eggs
All fresh fruit juices	Mayonnaise	Cranberries
Whole-grain breads	Onions	Cranberry juice
Chicken (not fried)	Gravies	Sanka
All fish (not fried)	Butter or margarine	Tea
Baked potatoes	Oils	Bananas
Salad	Chocolate	Lemonade and Acid drinks
Coffee	Red meat, pork, lamb	Sour cream
Rice	Salad dressing	Yogurt
All fruits except bananas	Custard	Mexican food
Plain cake and cookies	Milk, cream	Peanut butter
Jellied candies	Cream pies	Champagne
Marshmallows	Vegetables in the cabbage family	Red wine
Jello	Thick soups	Mozzarella or soft white cheeses
All green vegetables	Processed foods	
Jellies and jams	Honey	
Low-calorie drinks	Alcohol and beer	
Catsup	Ice cream	
Cereals	Cream cheese	
Chinese and Japanese food	Sharp yellow cheese	
White wine (noncarbonated)	Fried foods	
	Bacon, ham	
	Instant breakfast cereals	
	Hot dogs	
	Monosodium glutamate and artificial condiments	

Appendix C. Increasing Learning Potential: A Parent's Guide to Ancient Energy (Students 6-21)

1. Encourage students to meditate morning and evening. Arrange a quiet setting where meditation can take place. Daily practice helps clear the mind of distractions, allowing the more natural energies to enter the stream of consciousness.
2. Maintain a stress-free environment for family members, especially at mealtime. Avoid emotional confrontations that create a negative force field in the home.
3. Allow students to decorate an area of the home with personal effects, brightly colored posters, and other aesthetic touches.
4. Have books and magazines available for reading, but do not insist that the student read or force any type of learning activity that the student does not personally accept.
5. Provide daily outlets for discharging energies through exercise, work, social exchange, hobbies, and other activities.
6. Plan family menus with low-density foods. Avoid red meat and oils as much as possible in daily meal offerings.
7. Permit the playing of music in the home. Encourage the burning of brightly colored candles, incense, or other stimulants, while requiring that the student exercise reasonable caution in their use.
8. Discourage the use of narcotics, cigarettes, or alcohol, all of which suppress energy flow.
9. Do not overprogram the student. Allow ample free time for unstructured leisure. Encourage a positive attitude toward nature, plants, and animals.
10. Be aware of activities in the community that promote the

development of higher consciousness. Suggest that students attend these functions.

11. Encourage the development of spiritual principles without forcing the observance of religious ritual and doctrine.
12. Maintain contact with the school. Stay alert to parent activities where you have an opportunity to make changes in the educational program.
13. Foster creativity by encouraging the writing of poetry, the development of free-form art, and other unstructured pursuits. The more creative activities students are engaged in, the better channels they will become for the energies.
14. Demonstrate love and positive attitudes in the home. Observe student's actions and react to them in firm but positive ways.
15. Do not be surprised at anything young people say or do. If you cannot agree, refrain from criticism. Accept the philosophy of young people although you do not have to agree with it.
16. Make your own life a model for youngsters to follow. Do not preach morals—live them. Young people eventually will get the picture.
17. Discussions about illness should be deemphasized. Encourage students to envision their bodies as well. Do not "give energy" to illness, but rather work to bring the body back into alignment.
18. Verbal communication is not always essential. There are nonverbal ways to relate within the family group and for sharing personal thoughts with another person.
19. Have a sense of timing but do not impose your own rigid schedule on students. Be flexible and allow them to move at their own pace.
20. Know that learning can take place anywhere, at any time. When students are ready, they will learn what they need to know. Do not be impatient and do not judge your own children by the standards of others.
21. Encourage students to travel alone as early as possible. Provide a variety of different experiences from infancy on to stimulate diversity in thought and provide exposure to different force fields.
22. Whenever possible, allow students to have pets. Animals are good channels to encourage energy flow.
23. If you practice what you preach, the entire family will have unlimited energy all of the time. Do not force, coerce, criticize, or display negative emotions toward young people or they will react in kind. Maintain a positive environment and students will become natural channels for the flow of ancient energy.

Bibliography

Aguis, A. J. *The Hypogeum at Hal-Saflieni*. Rabat; Malta: Palbo Printing Press, ND.

Altered States of Awareness, Readings from Scientific American. San Francisco: W. H. Freeman, 1972.

"Ancient Egyptian Pyramid Thwarts Men of Science." *Los Angeles Herald Examiner*; December 8, 1974.

Asher, Maxine K. "Recent Theories of Intuitive Perception Applied to Ancient Anthropological Inquiry." Ph.D. dissertation, Walden University, July 1975.

______ *The Atlantis Conspiracy*. Los Angeles: Ancient Mediterranean Research Association, 1974.

Asimov, Isaac. *The Neutrino*. New York: Avon Books, 1966.

Barnett, Lincoln. *The Universe and Dr. Einstein*. New York: Bantam Books, 1968.

Bentov, Itzhak. *Stalking the Wild Pendulum*. New York: E. P. Dutton, 1977.

Bibby, Geoffrey. *The Testimony of the Spade*. New York: Alfred Knopf, 1956.

Blavatsky, Helene. *The Secret Doctrine*. Wheaton, Illinois: Theosophical Publishing House, 1962.

Bloom, Benjamin. *Taxonomy of Educational Objectives: The Affective Domain*. New York: Longman Green, 1956.

Bloomfield, Harold H., M.D.; Michael Peter Cain; Dennis T. Jaffe; and Robert B. Kary, *TM*. New York: Dell Publishing, 1975.

Brackman, Arnold. *The Dream of Troy*. New York: Mason and Lipscomb, 1974.

Brunton, Dr. Paul. *A Search in Ancient Egypt*. New York: Samuel Weiser, 1973.

Cameron, Verne T. *Map Dowsing*. Santa Barbara: El Cariso Publications, 1971.

Carter, Mary Ellen, and William A. M. C. McGarey. *Edgar Cayce on Healing*. New York: Paperback Library, 1972.

Castaneda, Carlos. *The Teachings of Don Juan: A Yaqui Way of Knowledge*. New York: Ballantine Books, 1971.

Cayce, Edgar Evans. *Edgar Cayce on Atlantis*. Virginia Beach, Virginia: Association for Research and Enlightenment, 1968.

Chadwick, John. *The Decipherment of Linear B*. 2nd ed. Cambridge: Cambridge University Press, 1970.

Chardin, Pierre Teilhard de. *Activation of Energy*. New York: Harcourt, Brace, Jovanovich, 1963.

——— *Human Energy*. New York: Harcourt, Brace Jovanovich, 1969.

Cohane, John Philip. *The Key*. New York: Crown Publishers, 1970.

Commoner, Barry. *The Closing Circle*. New York: Bantam Books, 1972.

Cox, Bill. "When Pyramid Tests Fail." *The Pyramid Guide*, Vol. III, No. 3, January-February, 1975, p. 1.

Emerson, J. N. "Intuitive Archaeology: The Argillite Carving." Unpublished Research Brief, Toronto, March, 1974.

Emiliani, Cesare, et al. "Paleoclimatological Analysis of Later Quaternary Cores from the Northeastern Gulf of Mexico." *Science*, Vol. 189, September 26, 1975, pp. 1083-1087.

——— "Intuitive Archaeology, A Psychic Approach." Toronto, March 1973 (unpublished).

Erman, Adolf. *The Ancient Egyptians*. New York: Harper & Row, 1966.

Fell, Barry. *America B.C.*. New York: QuadrangleNew York Times Book Co., 1976.

Ferguson, Marilyn. *The Brain Revolution*. New York: Taplinger Publishing, 1974.

Fuller, Buckminster R. *Intuition*. New York: Doubleday, 1972.

Garvin, Richard M. *The Crystal Skull*. New York: Doubleday, 1973.

Goodlad, John, and Frances M. Klein. *Looking Behind the Classroom Door*. 2nd ed. New York: C. A. Jones, 1974.

Goodman, Jeffrey. *Psychic Archaeology: Time Machine to the Past*. New York: G. P. Putnam's Sons, 1977.

______ *We Are the Earthquake Generation*. New York: Seaview Books, 1978.

Guilford, J. P. *The Nature of Human Intelligence*. New York: McGraw-Hill, 1967.

Hitching, Francis. *Earth Magic*. London: Cassell and Co., Ltd., 1976.

Hoover, Eleanor. "Atlantis-Anyone?" *Human Behavior*, November, 1973, pp. 10-12.

Jung, C. G. *Symbols of Transformation*. New York: Harper and Bros., Vols. I and II, 1956.

Karagulla, Shafica, M.D. *Breakthrough to Creativity*. Santa Monica: DeVorss, 1967.

Karlins, Marvin, and Lewis M. Andrews. *Biofeedback*. New York: Warner Books, 1975.

Koestler, Arthur. *The Roots of Coincidence*. New York: Vintage Books, 1972.

Kolosimo, Peter. *Timeless Earth*. London: Garnstone Press, 1973.

Kuhn, T. S. "The Structure of Scientific Revolutions: An Epitome." *Journal of the American Society for Psychical Research* 62 (1968).

Laxalt, Robert, and William Albert. "Land of the Ancient Basques." *National Geographic*, August 1968, Vol. 134, No. 2, pp. 240-277.

Leonard, George. *Education and Ecstasy*. New York: Delacorte Press, 1968.

Levi-Strauss, Claude. *The Structural Study of Myth*. New York: Basic Books, 1963.

Lilly, John C. *The Center of the Cyclone*. New York: The Julian Press, 1972.

Luce, Gay Gaer. *Biological Rhythms in Human and Animal Physiology*. New York: Dover Publications, 1971.

Marshack, Alexander. *The Roots of Civilization*. New York: McGraw-Hill, 1972.

Martineau, LaVan. *The Rocks Begin to Speak*. Las Vegas, Nevada: K. C. Publications, 1973.

Maslow, Abraham. *The Farther Reaches of Human Nature*. New York: Penguin Books, 1977.

Mclean, Adam. *The Standing Stones of the Lothians*. Edinburgh: Megalithic Research Publications, 1977.

Michell, John. *The Secret of the Stones*. New York: Ballantine Books, 1977.

Mitchell, Edgar D. *Psychic Exploration: A Challenge for Science*. New York: C. P. Putnams Sons, 1974.

Montessori, Maria. *Montessori Method*. New York: Schocken Books, 1964.

Moody, Raymond, Jr., M.D. *Life After Life*. New York: Bantam Books, 1977.

Moss, Thelma. *The Probability of the Impossible*. Los Angeles: J. P. Tarcher, 1974.

Moss, Thelma; John Ballard; and Alice F. Chang. "Time and ESP: An Unexpected Experiment." Los Angeles: N.D. (unpublished).

Nevill, W. E. *Geology and Ireland*. Dublin; Allen Figgis, 1972.

Nichols, R. Eugene. *The Science of Higher Sense Perception*. West Nyack, New York: Parker Publishing, 1973.

Ornstein, Robert E. *On the Experience of Time*. New York: Penguin Books, 1969.

——— *The Psychology of Consciousness*. New York: The Viking Press, 1972.

Ostrander, Sheila, and Lynn Schroeder. *Psychic Discoveries Behind the Iron Curtain*. New York: Bantam Books, 1973.

Parke, H. W. *A History of the Delphic Oracle*. Oxford: Blackwell Co., 1930.

Penfield, Wilder. *Mystery of the Mind*. Princeton: Princeton University Press, 1975.

Pillot, Gilbert. *El Codigo Secreto de la Odisea*. Madrid: Plazy Janes S.A., 1972.

Pomerance, Leon, F.S.A. *The Phaistos Disc*. Gotegorg, Sweden: Paul Astroms Forlag, 1976.

Ponte, Lowell. *The Cooling*. Englewood Cliffs, New Jersey: Prentice Hall, 1976.

Postle, Denis. *Fabric of the Universe*. New York: Crown Publishers, 1976.

Purce, Jill. *The Mystic Spiral*. New York: Avon Books, 1974.

Rayvenscroft, Trevor. *The Spear of Destiny*. New York: Bantam Books, 1974.

"Reaching Beyond the Rational." *Time* (International Edition), April 23, 1973.

Reich, Wilhelm. *The Cancer Biopathy*. New York: Noonday Farrar Strauss Giroux, 1973.

Renfrew, Colin. Before Civilization. New York: Alfred A. Knopf, 1973.

Rhine, Lousisa E. *Mind Over Matter*. New York: Collier Books, 1972.

Richardson, Alan. *An Introduction to the Mystical Qabalah*. New York: Samuel Weiser, 1974.

Ridley, Michael. *The Megalithic Art of the Maltese Islands*. Great Britain: Dolphin Press, 1976.

Sagan, Carl. *Other Worlds*. New York: Bantam Books, 1975.

Sapir, Edward. *Language*. New York: Harcourt Brace and Co., 1949.

Schanche, Don A. "Woman 'Recalls' Life with Pharoahs." *Los Angeles Times*, Wednesday, November 2, 1977. Part 1, pp. 16 & 17.

Schrader, Del. "Pyramid's Secrets Expected." *Los Angeles Herald Examiner*, June 13, 1976.

Schul, Bill. *The Psychic Frontiers of Medicine*. Greenwich: Fawcett Publications, 1977.

"Scuba Divers Report Finding Clues to Legendary Atlantis." *International Herald Tribune*, July 19, 1973.

Silberman, Charles. *Crisis in the Classroom*. New York: Random House, 1970.

Smith, Warren. *Lost City of the Ancients Unearthed*. New York: Zebra Books, 1976.

Spence, Lewis. *The History of Atlantis*. Secaucus, New Jersey: Citadel Press, 1973.

Schul, Bill. *The Psychic Power of Animals*. Greenwich, Conn.: Fawcett Publications, 1977.

Standing, E. M. *Maria Montessori: Her Life and Work*. New York: New American Library, 1962.

Steele, John. "The Bimini Road Complex." Unpublished Research Brief, London, 1977.

Steiger, Brad. *Medicine Power*. Garden City, New York: Doubleday, 1974.

Suares, Carlo. *The Cipher of Genesis*. New York: Bantam Books, 1973.

______ "Sunken Continent Discovered in Middle of Atlantic Ocean." *Miami News*, July 23, 1974.

Sykes, Egerton. *The Pyramids of Egypt*. London: Markham House Press, 1973.

Thompson, William Irwin. *At the Edge of History*. New York: Harper & Row, 1972.

______ *Time and Its Mysteries*. New York: Collier Books, 1962.

Toffler, Alvin. *Future Shock*. New York: Bantam Books, 1971.

"Tradition vs. Science—Indians Oppose Grave Excavations." *Roberts News-Marina News*, December 12, 1974.

Tomas, Andrew. *We Are Not the First*. New York: Bantam Books, 1973.

______ *Breaking the Time Barrier*. New York: Berkley Medallion Books, 1974.

Umland, Craig, and Eric Umland. *Mystery of the Ancients*. New York: Walker and Co., 1974.

Underwood, Guy. *The Pattern of the Past*. New York: Abelard-Schuman Ltd., 1973.

Valentine, J. Manson. "The Discovery and Possible Significance of X-Kulican, Ancient Mayan Site." *Alabama Museum of Natural History, Report 1*, February 1965.

Waters, Frank. *Masked Gods*. New York: Ballantine Books, 1975.

Watson, Lyall. *Supernature*. Garden City, New York: Anchor Press, Doubleday, 1973.

Watts, Alan. *In My Own Way—An Autobiography*. New York: Random House, 1973.

Van Daniken, Eric. *The Gold of the Gods*. New York: Bantam Books, 1974.

Weed, Joseph J. *Psychic Energy*. New York: Paperback Library, 1971.

Willoya, William, and Brown, Vinson, *Warriors of the Rainbow*. Healdsburg, Calif.: Nature Graph, 1962.

Zink, David. *The Stones of Atlantis*. New York: Prentice Hall, 1977.